From Broadcast.com
to the Future of Digital Media

STREAMING
WARS

An Insider's Journey from Broadcast.com
to Today's Global Streaming Empires

PATRICK SEAMAN

First edition, June 2025

Hardcover Edition Information:
    ISBN-13: 979-8-9878511-9-7
    Library of Congress Control Number: 2025937817

Paperback Edition Information:
    : ISBN-13: 979-8-9878511-8-0

Printed in the United States of America. Published by Milstar Books

This book is a work of nonfiction. The names, organizations, and events described herein are accurate to the best of the author's knowledge at the time of publication. Opinions expressed are solely those of the author and do not reflect the views of any companies or individuals mentioned.

Comparative Titles and Influences: Streaming Wars offers a firsthand perspective on the evolution of digital media, in the tradition of books like:

- *The Upstarts* by Brad Stone
- *That Will Never Work* by Marc Randolph
- *No Rules Rules: Netflix and the Culture of Reinvention* by Reed Hastings & Erin Meyer
- *How to Turn Down a Billion Dollars* by Billy Gallagher
- *The Facebook Effect* by David Kirkpatrick

These books explore the disruptive nature of internet-era innovation, streaming media, and entrepreneurial storytelling, paralleling the journey chronicled in Streaming Wars. For information about special discounts for bulk purchases, speaking engagements, or other inquiries, please contact: patrick@patrickseaman.com

Disclaimer

This book is intended for informational and educational purposes only. While every effort has been made to ensure the accuracy and timeliness of the information presented, the author makes no representations or warranties, express or implied, as to the completeness, reliability, or suitability of the content for any particular purpose.

The analysis, commentary, and company profiles included in this book are based on and solely arise from publicly available information, personal experience, and interviews conducted at the time of writing. These sources include corporate websites, public case studies, media coverage, investor briefings, and third-party research platforms. No proprietary data was used, and no company profiled was given the opportunity to review or verify their entry. As such, offerings, pricing models, technologies, and organizational strategies may have evolved since publication.

Nothing in this book should be construed as investment advice, legal guidance, or a commercial endorsement. Readers are encouraged to conduct their own independent research and consult appropriate professionals before making business, technical, or financial decisions.

The views expressed are solely those of the author and do not necessarily reflect the opinions of any affiliated organizations, companies, or collaborators. Any opinions or interpretations relating to historical events, including those in which the author was personally involved, are the author's alone and are not intended to represent the views or positions of any other individuals or entities, past or present.

DEDICATION

To the founders and team at AudioNet.com / Broadcast.com:

You were pioneers in every sense of the word.

Before the rest of the world saw what streaming could be, you built it. You coded, hacked, scaled, sold, travelled the globe with me, put up with me, and streamed when the infrastructure barely existed and the vision was years ahead of its time.

You all lit the fuse and strapped yourselves to that rocket ride together!

This book stands on your shoulders.

Thank you for the late nights, the wild experiments, the crazy optimism, and the belief that broadcasting could be democratized: not just by bandwidth, but by ambition.

You changed the game. And we're still playing.

■ Patrick Seaman

PATRICK SEAMAN

Introduction: The Buffer Wheel

Dallas, Texas - March 2025

My phone buzzed with a notification while I was grabbing coffee. "Your live game is about to start!"

I tapped the alert, and instantly, high-definition video filled my screen - the crowd roaring, players taking the field, stats and betting odds floating elegantly at the bottom of the display. No buffering, no lag, crystal clear even over cellular. I slipped in my earbuds, selected the commentary track I wanted, and just like that, I was connected to a stadium thousands of miles away.

All of this happened in seconds, and none of it felt remarkable.

But it should.

Because less than thirty years ago, I was in a cramped room on Elm Street in Dallas, surrounded by repurposed cafeteria tables bowing under the weight of humming desktop computers, watching as the glacially slow progress bar of RealAudio crept forward, desperately trying to stream a college football game online for the very first time.

The year was 1995, and the bandwidth for that pioneering broadcast was roughly one-hundredth of what your smart refrigerator uses today.

If you ask me what it was like in those early days of streaming media, I suppose I could compare it to the Wright brothers at Kitty Hawk. We weren't building jetliners. We were trying to make something, anything, get off the ground, even if just for a few seconds.

The first flights were short, unstable, and hardly impressive by today's standards. But they proved something crucial: it could be done.

That first streaming flight changed everything.

Today, we take for granted platforms like Netflix, YouTube, and Twitch. We expect instant, high-definition content anywhere, anytime. We can't imagine waiting more than a few seconds for a video to start or tolerating the dreaded buffer wheel for more than a moment.

But the buffer wheel – that spinning icon of digital patience – was once our constant companion, a reminder that we were pushing the very limits of what networks could handle.

This book isn't just about the pioneering days of streaming, though those wild early experiments form its foundation. It's a comprehensive journey from the first stuttering RealAudio streams to today's 4K HDR content delivered instantly to billions of devices worldwide.

Between these pages, you'll discover:

- How the technology evolved from dial-up modems to fiber optics and 5G, and why you still experience frustrating delays during live sports broadcasts
- The high-stakes battles for content rights that transformed billion-dollar industries and made stars out of everyday creators
- Why the digital media you "purchase" isn't really yours at all, and how companies can – and do – alter or remove content long after you've paid for it
- The hidden influence of algorithms that decide what content succeeds, what fails, and what you'll likely watch next
- How audience rebellions, from "ratio-ing" to mass social media campaigns, are wresting control from traditional media gatekeepers
- The next frontier of sports streaming that could finally solve the stadium connectivity challenge

From AudioNet's jerry-rigged server rooms to YouTube's algorithm-driven empire, from the birth of Netflix to the decline of cable TV, from the explosion of user-generated content to the rise of AI-generated media, this is the complete story of how we arrived at today's streaming landscape – and where we're headed next.

I've witnessed the entire evolution of streaming media, from those first shaky broadcasts at AudioNet (which became Broadcast.com) to today's immersive, real-time experiences. I've seen billion-dollar empires rise and once-dominant platforms fall. I've watched as the power shifted from traditional gatekeepers to tech giants to everyday content creators and now to algorithms and the audience itself.

The streaming wars didn't end with Netflix's victory over cable or YouTube's dominance in user-generated content. They're still raging, with new fronts opening in sports, gaming, music, and beyond. And the final winners haven't been decided.

So whether you remember the screeching sound of dial-up modems connecting to the internet or you've never known a world without high-speed WiFi, this journey matters to you. Because to understand where media is going – and who will control what you see, hear, and experience next – you need to understand how we got here.

And it all started with that buffer wheel, slowly turning, promising that something amazing was about to happen....

PATRICK SEAMAN

1: The Dawn of Online Streaming – AudioNet & Broadcast.com (1995-1999)

How We Built the First Streaming Empire – And What We Learned

---

"I had to climb through a second-story window with a broken ladder to keep the streams running."

That's not a metaphor - it's literally how we built the streaming industry in 1995. Back then, – and video over the internet wasn't just ambitious - it was considered impossible. With dial-up modems screeching at 56K and websites that were mostly text and low-resolution images, we were trying to do something that shouldn't have worked.

But it did. This is the story of how AudioNet, which became Broadcast.com, pioneered an industry that would eventually transform how the world consumes media.

In 1995, – and video over the internet sounded like science fiction. That year, I helped launch **AudioNet**, a company that aimed to **bring**

live radio and video to the internet. The original business plan was to take the color sports commentary from five radio stations and post the audio online.

For readers who never experienced the internet in 1995, imagine this: You sit at your computer and click "connect," triggering a cacophony of electronic screeching, static, and beeps as your modem negotiates a connection with your internet provider. This process could take 30-60 seconds, and any interruption meant starting over. Once connected, your 56K modem (the fastest commonly available) transferred data at roughly 5 kilobytes per second – about 0.005% of today's average broadband speed. A single MP3 song might take 10-15 minutes to download. Loading a basic webpage with a few images could take 30 seconds. And if someone picked up the phone elsewhere in the house? Instant disconnection. This was the technological reality we were trying to stream audio through – the equivalent of attempting to push a firehose worth of water through a drinking straw.

It wasn't even live.

We had to spend hours encoding audio files just so people could listen to them after the fact.

The first shows we streamed on the internet were *David Gold*, *Wally Lynn*, *Leon Simon*, and *Norm Hitzges*, all of which aired on **KLIF Radio**. It was AudioNet that put KLIF online, making it the first commercial radio station on the internet.

Back then, the process was painstakingly slow, sometimes taking hours for every hour encoded. The software was also notoriously "flaky," prone to crashing mid-process, forcing us to start over and lose valuable time. Despite these challenges, the technology quickly evolved, and so did our business plan as we navigated numerous pivots and grappled with the explosive growth that began from day one. Of course, it didn't help that the computers we could afford to buy were… not great. Packard Bell or worse. We made the most of it, though.

When Mark Cuban and Todd Wagner hired me, it was to consult for 10 hours a week. Within a month, it was 40. By two months, it was 80 and beyond. Realizing this wasn't sustainable, Todd decided to formalize my role as their first official, actual employee. Eventually, the rocket ride turned into **Broadcast.com** and became the **largest aggregator and distributor of audio and video in the world** by 1999.

When Broadcast.com went public in July 1998, the response from Wall Street was extraordinary. The stock surged an unprecedented 250% on its very first trading day, marking one of the most dramatic IPOs in internet history. Overnight, the company's valuation reached approximately $1 billion, instantly catapulting Mark Cuban's net worth to around $300 million and Todd Wagner's to roughly $170 million. This remarkable debut not only cemented Broadcast.com's position as a leader in the burgeoning streaming industry, but also signaled a seismic shift in the market's perception of internet-driven businesses.

Before it was AudioNet, by the way, as I recall, (Cameron) Chris Jaeb and Todd Wagner approached Mark about a way to enable access to out-of-town sports games. It was originally, for a brief time, "Cameron Broadcast Network."

Mark Cuban immediately recognized the potential and invested $10,000 for an initial 2% stake in the company. Soon after, Cuban and Wagner structured a pivotal deal with Jaeb. Chris retained a 10% equity stake along with a modest monthly salary of $2,500, while Cuban took operational control, setting the stage for the aggressive growth that followed.

I set up equipment at PicNet (a local ISP), at 2929 Elm Street, and at Mark's house. The servers crashed a lot, so I ended up going back and forth between all three at all hours of the day and night. In those days, Mark intentionally never locked his door. When he was out of town, on more than one occasion, I had to break into his house – because his brother had come by and locked the door in Mark's absence. There was a ladder in his back yard with one side missing. I would use it to climb up to an unlatched window on the 2nd floor.

Radio Shack

The basic initial deal we made with radio stations was, in short, to 'put them on the internet' in exchange for some of their unsold ad spots, which we would then aggregate and sell nationally.

Most radio stations did not have internet access at the time, or if they did, it was ISDN for audio connections. If we'd had to pay to give them an internet connection, it would have been expensive in terms of both time and money to set up. Remember, this is the mid 1990s. Also, they often wouldn't even have the rack space available.

So, what we would do is set up an encoder(s) at a local internet service provider and connect a radio to the encoder. Often, we had to run a wire to the roof to get a signal.

The radios we used at first had manual analog tuners that would eventually drift when left unattended for long periods of time. I decided to look for an inexpensive digital radio we could switch to.

Radio shack had $100 digital radios with a 1/8th jack.

Many of the cities we went to had more than one station we put online. My SOP was to go to the local Radio Shack and buy the (exactly 2) digital radios they would have in stock. After a year of doing this across the country, my suspicion was that, apparently, Radio Shack thought that there was a huge demand. They started having a bunch of the radios at their stores, but by then, we didn't need them anymore as our tech was evolving. I felt a little bad about that.

Of course, I'd be remiss if I didn't acknowledge just how often the presence of a local Radio Shack saved us during those early days.

Truthfully, we counted ourselves lucky that they seemed to have stores in nearly every city we visited. Our road kits took quite a while to evolve into the comprehensive collections of gear they eventually became, and it always felt like something critical was missing. Whether it was a specialized cable, an unusual connector, or a last-minute adapter to resolve unexpected technical challenges at live events, Radio Shack often bailed us out. And it wasn't just live events: as we began to install and set up more stations, spontaneous requests would pop up – something like, "Hey, while you're already out there, could you swing by XYZ and get them online too?" I'm sure you get the idea; those little stores played a surprisingly big role in keeping us operational.

Eventually, our Pelican and Anvil cases were – at least most of the time – properly stocked and ready to go. Gradually, professional-grade gear began replacing the consumer-level equipment we'd initially relied upon, which significantly improved our reliability. But keep in mind that when you're constantly packing up delicate electronics,  transporting them repeatedly to new venues, and subjecting them to the rough handling and bouncing around inherent in shipping and travel, things inevitably wear out and fail. To combat this, we established a strict rule, one I've since repeated many times elsewhere: whenever a cable stopped working or became unreliable, we immediately cut it in half and threw it away. We simply couldn't risk putting a faulty cable back into the Pelican case, only to unknowingly reuse it later at another event and then waste precious time chasing down the same problem all over again.

Live events didn't just wear down cables and gear – they took their toll on people, too.

Modem Technology

The modems powering these early streaming attempts were marvels of compromise engineering. These devices – whose name derives from "modulator-demodulator" – converted digital data from computers into analog signals that could travel across standard telephone lines, then back to digital at the receiving end. They worked by translating binary computer data into audible tones, essentially "singing" information across phone lines designed for human voices. Early streaming required specialized protocols like UDP (User Datagram Protocol) because traditional TCP connections would simply stall when packets were lost. Instead, our streams would keep moving forward, sacrificing perfect fidelity for continuity – better to have occasional audio glitches than constant buffering. These technical workarounds were essential but imperfect solutions, creating the characteristic stuttering and quality fluctuations early listeners came to expect as part of the streaming experience.

Early Audio Codecs

During the mid-1990s, several notable audio codecs emerged, laying the groundwork for the streaming revolution. Among the pioneers were Xing Technologies and RealAudio, but they were not alone in shaping the audio landscape of the internet's early days.

Xing Technologies: Known for being one of the early adopters and implementers of the MP3 format, Xing Technologies developed an encoder/decoder system that prioritized speed. However, while its encoding speed was impressive, it was often criticized for producing lower audio quality compared to competitors. Additionally, Xing required proprietary hardware, significantly driving up costs for users and limiting its adoption to those who could afford the high price tag.

We ultimately decided to not go with Xing. It was just too expensive and proprietary and would have made us too dependent upon them for both encoding software and hardware. If all we were doing was "a" stream, it might have been okay. That wasn't for us. We knew we were going to need a scalable platform that could handle our rapidly growing needs.

RealAudio: Introduced by RealNetworks in April 1995, RealAudio was groundbreaking for its ability to stream audio over the internet. It was a significant technological leap at the time. It used a variety of codecs to accommodate different bitrates and levels of audio fidelity, making it adaptable for varying connection speeds, from dial-up to faster connections. RealAudio quickly became a de facto standard for streaming in the late '90s, offering a relatively seamless experience for users compared to other emerging technologies.

We quickly adopted Real as our software platform of choice in 1995. This left us free to build out the hardware side using more cost-effective solutions.

The brainchild of Rob Glaser, a former Microsoft executive, Rob leveraged his deep understanding of software and networking to position RealAudio as a leader in the nascent streaming industry. Originally named Progressive Networks, the company became RealNetworks.

Glaser's departure from Microsoft was notable not just for the timing but also for the tension it introduced between the two companies. Glaser had seen the potential of media on the internet early, and RealNetworks capitalized on this foresight by releasing RealAudio, which allowed users to stream content even on limited bandwidth connections.

Windows Media Player: Microsoft first introduced its Windows Media Player as a streaming media platform with the release of Windows Media Player 6.1 in 1998. Microsoft's entry into the streaming market marked the beginning of a rivalry with RealNetworks. While RealAudio retained its dominance due to its early market entry and broad adoption, Microsoft's backing of Windows Media Player and its integration into Windows operating systems presented a significant challenge.

RealNetworks accused Microsoft of anticompetitive behavior, culminating in lawsuits that highlighted the struggles of smaller innovators facing tech giants. But more on the rivalry and the ongoing serious problems we had with the RealNetworks platform, later in this series.

Despite these challenges, RealAudio's adaptability and early leadership in the field made it indispensable during the first wave of Internet media. The software became widely used by radio stations, educational institutions, and businesses to deliver audio content, setting the stage for future innovations in streaming.

Others worth mentioning:

- **Voxware**: A lesser-known but notable competitor, Voxware specialized in speech compression and was used in applications where bandwidth was extremely limited. Its codecs were particularly tailored for voice transmission rather than music, finding niche applications in early VoIP and educational tools.

- **Fraunhofer IIS**: The institute responsible for the **MP3** format itself, Fraunhofer's encoder, was regarded as the gold standard for MP3 quality during this era. Although less accessible to consumers than RealAudio, it was widely used in professional applications and laid the foundation for MP3's dominance in digital music.

- **QDesign Music Codec**: Known for its use in Apple's **QuickTime**, the QDesign codec provided high-quality audio at lower bitrates, making it an efficient choice for early multimedia applications. Although less prominent than RealAudio or Xing, it had a dedicated user base.

- **TwinVQ** (Transform-domain Weighted Interleave Vector Quantization): Developed by NTT in Japan, TwinVQ was another audio codec that focused on delivering high-quality audio at low bitrates. While it never gained widespread traction, it represented the global interest in optimizing audio compression.

Conclusion

These early technical battles - choosing codecs, jerry-rigging Radio Shack equipment, and climbing through windows to reboot servers - laid the foundation for today's streaming giants. But the real revolution was only just beginning. The solutions we cobbled together in those early days of AudioNet have evolved into technologies that power everything from Netflix's global content delivery network to Spotify's personalized music recommendations to video calls with family across the world. What began as experimental broadcasts to a few hundred listeners now enables billions of streaming sessions daily across devices we couldn't have imagined in 1995.

While the technology has advanced exponentially, many of the core questions remain the same: How do we deliver content efficiently? Who owns the rights? And most importantly, how do we connect people to the entertainment and information they value most? As we moved into 1996, we were about to discover that getting the technology to work was only the first challenge. The real test would be scaling it to serve hundreds of thousands of simultaneous listeners - something no one had ever done before.

Bandwidth: Then -vs- Now

Consider how far streaming technology has evolved since those early days:

Evolution of Streaming Bandwidth: 1995-2025

Metric	1995	2025
Typical stream bitrate	8-28 Kbps (mono)	5-20 Mbps (4K video with surround sound)
Time to buffer 10 seconds of content	~30+ seconds	Instantaneous
Maximum concurrent listeners on AudioNet/Broadcast.com (1999)	~50,000+ during major events	Netflix handles 100+ million simultaneous streams
Audio quality	AM radio to FM radio equivalent	CD-quality to lossless high-fidelity
Cost of bandwidth	$10,000+ per month for enough capacity to stream to a few thousand listeners	Fractions of a penny per gigabyte
Reliability	Frequent disconnections and buffering	99.9% uptime expected

What we achieved with a handful of talented engineers and hundreds of thousands in equipment can now be accomplished with a smartphone and a basic streaming app. The underlying principles remain similar, but the scale, quality, and accessibility have been transformed beyond what we could have imagined.

Early Competition and Failed Attempts in the Streaming Landscape

While we were building AudioNet, we weren't alone trying to crack the streaming puzzle. The digital frontier attracted numerous pioneers, though most eventually faded or pivoted as the technical and business challenges proved overwhelming.

Progressive Networks (later RealNetworks) focused primarily on building streaming technology rather than aggregating content as we did. Rob Glaser, a former Microsoft executive, created a vertically integrated approach – owning the entire tech stack from servers to players. This strategy gave them tight control but limited their flexibility and scale. While their RealPlayer became ubiquitous, they struggled to decide whether they wanted to be a platform company or a broadcaster, an internal schism that ultimately hampered their growth.

Xing Technology took a different approach with their premium hardware encoders. They targeted high-end corporate users with impressive quality but missed the consumer market entirely due to prohibitive costs. Their specialized hardware approach created beautiful results but couldn't scale to meet the explosive growth of internet adoption.

PseudoTV, launched in 1995 with former MTV VJ Adam Curry at the helm, attempted to become the "MTV of the internet" but struggled with the bandwidth limitations of the time. Their ambitious vision of full-motion video streaming was simply too far ahead of what dial-up connections could reliably deliver, and they eventually pivoted away from streaming altogether. NetRadio, another early competitor, created internet-only radio stations but lacked the live events and sports content that drove our rapid growth. We actually acquired them later on, absorbing their programming into our expanding content library.

Several telecommunications companies also attempted streaming platforms but approached the challenge as a network problem rather than a content opportunity. They had the infrastructure but not the programming people wanted – the equivalent of building highways without having any cars to drive on them. Meanwhile, traditional media companies initially viewed streaming as a threat rather than an opportunity, leaving the field open for startups like ours to establish a foothold.

What set AudioNet/Broadcast.com apart was our hybrid approach – we understood that both technology and content acquisition were equally critical, and we remained nimble enough to adapt as the landscape evolved. Unlike tech-only companies, we focused intensely on securing compelling content that audiences couldn't get elsewhere. Unlike content-only companies, we invested heavily in building infrastructure that could actually deliver reliable streams. While others tried to force existing business models onto this new medium, we were willing to invent new approaches that suited the internet's unique characteristics.

Most importantly, we recognized early on that live events, particularly sports, created a sense of urgency that drove adoption despite the technical limitations of early streaming. These efforts were buttressed by our live and on-demand corporate events. Intel, in particular, with their ambitious live events in the US and Asia, saved our skins and our bottom line over and over again during the first few years. When most streaming companies were still figuring out their business models, we were already generating revenue through these corporate partnerships.

By 1998, the graveyard of failed streaming pioneers was already filling up, but their collective experiments helped define what worked and what didn't in this emerging medium. Silicon Valley threw mountains of venture capital at California streaming startups that garnered splashy press coverage in Wired and Red Herring. Meanwhile, we trucked along in Dallas, taking advantage of our lower operating costs, Texas work ethic, and stubborn determination to make the technology work regardless of the obstacles.

We learned from competitors' missteps as much as from our own successes, continuously refining our approach as the technology and market evolved. This willingness to adapt, combined with Mark Cuban's sales prowess and our focus on securing exclusive content rights that nobody else initially valued, positioned AudioNet/Broadcast.com to dominate the early streaming landscape when others couldn't figure out how to turn technological possibility into a viable business.

Where they are now

Mark Cuban – After selling Broadcast.com to Yahoo! for $5.7 billion in 1999, Cuban shrewdly sold his Yahoo! stock before the dot-com crash, preserving his billionaire status. He purchased the Dallas Mavericks in 2000 for $285 million (now valued at over $4 billion) and led them to an NBA Championship in 2011. Cuban became a household name as a "shark" investor on ABC's "Shark Tank" from 2011-2023, investing in hundreds of startups. Today, he remains actively involved in technology investments, healthcare initiatives through his Cost Plus Drugs company, and continues to be an outspoken figure in business, sports, and occasional political commentary.

Todd Wagner – Following the Yahoo! acquisition, Wagner partnered with Cuban to form 2929 Entertainment, which acquired Landmark Theatres, Magnolia Pictures, and the HDNet (now AXS TV) cable networks. He later founded the Charity Network, including Charitybuzz and Prizeo, digital fundraising platforms that have raised over $500 million for nonprofits. Wagner has focused increasingly on philanthropic efforts through the Todd Wagner Foundation, which supports educational and entrepreneurial programs for underserved communities.

Chris (Cameron) Jaeb – After departing AudioNet with his 10% stake, Jaeb founded several companies including GoodBiz113 and Celebrate Express. He later embraced sustainable entrepreneurship, focusing on ventures promoting environmental and social responsibility. Jaeb has become an advocate for conscious capitalism and social entrepreneurship, investing in and advising mission-driven startups.

Belinda Johnson – Johnson, our in-house counsel who helped navigate the complex legal challenges of early streaming rights, went on to an impressive career in tech. She served as Airbnb's first Chief Operating Officer from 2018-2020 after joining as Chief Business Affairs and Legal Officer in 2011. Prior to Airbnb, she held senior legal positions at Yahoo! following the Broadcast.com acquisition. Johnson has served on several corporate boards including PayPal and continues to be a respected voice on technology law and business operations.

RealNetworks/Rob Glaser – RealNetworks, once dominant in streaming media, struggled to maintain relevance as newer technologies emerged. The company diversified into gaming, mobile apps, and AI-driven services. Rob Glaser, who stepped down as CEO in 2020 (after an earlier departure and return), remains Chairman of the

Board. Despite introducing innovations like RealPlayer, the company never achieved the scale of later streaming giants, with its stock price declining significantly from its late-1990s peak.

Jim Cline – My friend Jim Cline, who helped pioneer internet audiobooks through our collaboration on "A Small Percentage," passed away in 2024. A graduate of Virginia Tech (1974-1978) and the University of Kentucky (M.S. in Electrical Engineering, 1979-1980), Jim combined his technical expertise with remarkable creative vision. We co-authored "Website Sound," published by New Riders/McMillan in 1997/1998, one of the first comprehensive guides to audio implementation on the early web. Later, we co-founded Timberwolf Press, which produced award-winning full-cast audio productions that set new standards for audio entertainment in the digital age. Jim's innovative approach to digital storytelling helped establish formats that would later become commonplace in the podcast era. His unique ability to bridge engineering precision with artistic creativity made him an invaluable collaborator during the early days of streaming and beyond.

In Chapter 2

How we became the world's largest streaming platform and pioneered live sports broadcasting on the internet.

What are your earliest memories of streaming media? Were you one of those early AudioNet listeners trying to tune in through a dial-up connection? Or maybe you remember the first time you realized you could listen to a radio station from across the country?

And if you found this interesting, in Chapter 2 I'll reveal how we scaled from those makeshift Radio Shack setups to becoming the largest streaming platform in the world.

2: Before YouTube, Before Netflix

There Was Broadcast.com

Bridging the Broadcast Era to the YouTube Revolution

I f the 1990s were about proving streaming was possible, the 2000s were about proving it was inevitable.

When we started AudioNet, later Broadcast.com, the idea of watching or listening to live content over the internet felt like science fiction. We had to jury-rig radios, climb through windows to reboot servers, and make do with early codecs that barely worked. But despite the technological duct tape holding it all together, we built something that changed how people consumed media.

By 1999, Broadcast.com had become the largest streaming platform in the world and, arguably, the largest aggregator and distributor of audio and video in the world. Then Yahoo! bought us, fumbled the ball, and left an open field for whoever could pick it up next.

On April 1, 1999, Yahoo! acquired Broadcast.com for an astonishing $5.7 billion, marking Yahoo!'s most expensive acquisition ever. For valuation purposes, the official user count was set at 570,000

– a financial metric necessary for the deal, though our major live events routinely attracted audiences far exceeding that figure. Perhaps that was registered users? I don't recall. Anyway, this official number effectively valued each user at around $10,000, an extraordinary valuation that vividly captured the frenzy and boundless optimism of the dot-com boom. Shortly after the acquisition, Mark Cuban swiftly sold his Yahoo! shares, netting over $1 billion, a decisive move that highlighted his sharp business instincts and underscored the immense volatility characteristic of internet valuations during that era.

Yahoo's Missed Opportunity: How a Tech Giant Fumbled Streaming's Future

Yahoo's acquisition of Broadcast.com should have positioned them to dominate the emerging streaming landscape, but a series of strategic missteps squandered this potential. First, there was a fundamental culture clash. Yahoo was primarily a portal and search company with an advertising-driven business model, while Broadcast.com was built on content partnerships and subscription services. Instead of preserving Broadcast.com's unique business approach, Yahoo attempted to force it into their existing advertising framework.

After the pooling of assets legal two-step, which, by the way, bypassed Audionet / Broadcast.com employee incentive stock agreements that would have vested shares in the event of a merger, and before I left in 1999, I visited Yahoo in an attempt to address prospective synergies. I was rebuffed. Yahoo had no real interest in hearing from me or the team that had build the largest aggregator and distributor of audio and video in the world. They were Yahoo, after all. We were just another check mark on the acquisitions todo list.

The technical integration was equally problematic. Rather than maintaining Broadcast.com's nimble, dedicated streaming infrastructure, Yahoo migrated the services to their more generalized systems, sacrificing performance and reliability in the process.

Engineers who had built Broadcast.com's innovative systems from the ground up found themselves fired, sidelined or reassigned to unrelated projects. By 2002, nearly all of Broadcast.com's original technical team had departed.

Perhaps most damaging was Yahoo's failure to recognize streaming's strategic importance. While companies like YouTube (founded in 2005) and Netflix (which began streaming in 2007) were building their entire business models around streaming technology, Yahoo relegated it to just another feature in their sprawling portal. The specialized knowledge and technology that could have made Yahoo the dominant streaming platform was instead diluted across dozens of different Yahoo properties.

The final blow came from Yahoo's lack of content strategy. We had established relationships with sports leagues, radio stations, and television networks that Yahoo largely neglected. These partnerships withered as rights expired and weren't renewed. By 2002, most of Broadcast.com's original content partnerships had expired or moved to competitors, leaving Yahoo with little to stream despite owning the platform that could have delivered it.

The cautionary tale is clear: technology alone isn't enough. Without the right organizational structure, business model, and content strategy, even groundbreaking innovations can wither. Yahoo had purchased a Ferrari but treated it like just another vehicle in their fleet, failing to recognize its unique capabilities until competitors had built newer, better models.

By 2006, when Google acquired YouTube for $1.65 billion (less than one-third of what Yahoo paid for Broadcast.com), Yahoo's streaming infrastructure had been largely dismantled or repurposed. The company that could have dominated video streaming was instead struggling to remain relevant in a landscape now shaped by the very innovations they had failed to nurture.

Streaming Media: Then and Now

Streaming at Scale: 1999 -vs- 2025

Metric	Broadcast.com (1999)	Streaming Giants (2025)
Peak Concurrent Viewers	~500,000 (Super Bowl)	200+ million (YouTube World Cup)
Video Quality	240p at best, often smaller	4K HDR standard, up to 8K available
Bandwidth Required	28-56 Kbps for audio,300 Kbps for video	5-15 Mbps for HD,25+ Mbps for 4K
Buffering Time	30+ seconds common	Near-instantaneous
Content Library	65,000 hours of on-demand programming	Netflix: 100,000+ hours, YouTube: 800+ million videos
Global Reach	Primarily North America with limited international	190+ countries standard
Infrastructure Cost	$10,000+ per server, $100,000+ monthly bandwidth	Pennies per GB via cloud infrastructure
Mobile Accessibility	None (pre-smartphone era)	70%+ of streaming happens on mobile devices
Revenue Model	Primarily advertising and corporate services	Subscriptions, ads, microtransactions, memberships
Content Creation	Professional studios only	From smartphones to Hollywood productions
Delivery Method	Direct server connections	Complex CDNs with edge computing
Staff Required	Dozens of engineers for basic reliability	AI-managed systems with minimal human intervention

PERSPECTIVE: The bandwidth required to stream a single 4K video today could have supported more than 1,000 simultaneous audio streams on Broadcast.com in 1999. What once required rooms full of servers can now be accomplished with a single smartphone. The amount of video uploaded to YouTube in a single day exceeds what Broadcast.com could stream in its entire existence. Yet despite these dramatic advances in scale and technology, the core principles remain the same: delivering compelling content to audiences when and where they want it.

The Dot-Com Crash: Streaming's Winter and Unexpected Renaissance

The dot-com crash of 2000-2001 hit the streaming industry particularly hard, creating a barren landscape where once-promising startups withered and investment dollars evaporated almost overnight. Yahoo's stock, which had peaked above $118 in January 2000, plummeted to below $9 by September 2001, dramatically reducing the actual value of the Broadcast.com acquisition. The crash wasn't just a financial correction – it was an existential crisis for the entire streaming ecosystem.

In the immediate aftermath, streaming innovation appeared to stall. Dozens of Broadcast.com competitors and adjacent streaming services shuttered completely. Those that survived made drastic cuts to engineering teams and R&D budgets. Bandwidth costs remained prohibitively high, and without the cushion of venture capital, experiments in live streaming and interactive content were largely abandoned. Between 2001 and 2004, it seemed the streaming revolution might be postponed indefinitely.

Yet paradoxically, this market correction created the conditions that would ultimately enable streaming's renaissance. Three critical developments emerged from the crash:

First, infrastructure costs finally began to drop. The massive overbuilding of fiber-optic networks during the bubble years created a bandwidth glut that, while devastating for telecommunications companies, gradually made high-quality streaming economically viable for new entrants.

Second, the crash cleared away unsustainable business models. The survivors were forced to focus on genuine value creation rather than speculative growth. This reset allowed for more measured, sustainable approaches when the industry revived.

Third, and perhaps most importantly, the vacuum created by retreating corporate giants left space for entrepreneurial innovation from unexpected quarters. YouTube, founded in 2005 by three former PayPal employees, faced much less competition than it would have encountered years earlier. Similarly, when Netflix began experimenting with streaming in 2007, the major media companies were still too shell-shocked from the crash to mount serious competition.

The crash also redirected talent and expertise. Engineers and executives who had cut their teeth at Broadcast.com and similar pioneers didn't abandon the field – they brought their knowledge to new ventures, often with more sustainable business models. Many pivotal figures at YouTube, Netflix, and later streaming platforms had previously worked at companies that didn't survive the crash.

In many ways, the dot-com crash functioned like a forest fire – devastating in the short term but ultimately clearing deadwood and enriching the soil for new growth. When streaming finally reached mass adoption in the late 2000s, it did so with more efficient technologies, more realistic business models, and a clearer understanding of what audiences actually wanted from the medium.

The boom-and-bust cycle created a five-year gap between Broadcast.com's innovations and their mainstream adoption through new platforms. But this enforced patience allowed the technologies to mature, bandwidth costs to drop, and consumer expectations to develop. By the time YouTube and Netflix Streaming emerged, the foundation was laid for sustainable growth rather than another speculative bubble.

9/11's Shadow: How Tragedy Reshaped Media and Streaming

The aftermath of the dot-com crash was still unfolding when the September 11, 2001 terrorist attacks delivered another profound shock to the emerging digital media landscape. The tragedy's impact on the industry was both immediate and far-reaching, affecting everything from business deals to content distribution to consumer behavior.

After broadcast.com, at Timberwolf Press, I experienced this disruption firsthand. We were set to close a deal to acquire a competitor on September 11 itself – a transaction that would have expanded our audiobook catalog and production capabilities. When the attacks occurred, everything stopped. The deal, like countless others across the country, evaporated in the wave of uncertainty that followed. But the repercussions went far beyond canceled acquisitions.

In the months after 9/11, our orders virtually disappeared. Libraries, which had been reliable customers for our audiobooks, froze their purchasing for nearly a year as their budgets were redirected or cut. This pattern played out across the publishing and entertainment sectors, with the American Booksellers Association reporting that independent bookstore sales dropped by 15-20% in the weeks following the attacks, and remained depressed for months afterward.

For the broader streaming industry, which was already reeling from the dot-com crash, 9/11 created a complicated set of challenges. On one hand, internet traffic surged as Americans sought real-time news and information, with CNN.com reporting a 1,300% increase in traffic on September 11. The tragedy demonstrated the internet's vital role as an information source during crises, but simultaneously exposed critical weaknesses in content delivery infrastructure. Major news sites crashed under the unprecedented load, highlighting how unprepared early streaming platforms were for mass concurrent usage.

The aftermath of 9/11 also transformed media consumption patterns in ways that would ultimately shape streaming's evolution. A Nielsen report from October 2001 showed television viewership spiking by 33% in the weeks following the attacks, with news channels seeing the largest gains. This renewed focus on real-time information created both opportunities and complications for digital media:

- Demand for live streaming news increased dramatically, but bandwidth constraints and technical limitations prevented most providers from capitalizing on this interest
- Consumers developed new expectations for immediate access to breaking news, accelerating the shift away from printed newspapers and toward digital sources

- Media consolidation accelerated as smaller companies struggled to survive the combined economic impacts of the dot-com crash and post-9/11 advertising downturn

The advertising market, which underpinned much of online media, contracted sharply, with digital ad spending declining for the first time in its history. According to the Interactive Advertising Bureau, online advertising revenue dropped nearly 12% in the first nine months following 9/11. For streaming startups already operating on thin margins, this revenue collapse was often fatal.

Media giants weren't immune either. Viacom (owner of MTV, CBS, and Paramount) reported a 33% drop in profits for the quarter following the attacks. Disney saw attendance at its parks plummet while its ABC network incurred massive costs from commercial-free news coverage. These financial pressures meant that major media companies, which might otherwise have been investing in digital initiatives, instead focused on preserving their core businesses.

Perhaps most significant for the streaming industry was the shift in investor sentiment. Venture capital, already scarce after the dot-com crash, became even more risk-averse. According to PricewaterhouseCoopers, VC investments in media and entertainment fell by more than 60% in the year following 9/11. Funding for streaming innovations wouldn't recover to pre-crash levels until nearly 2006.

The combined effect of these economic factors delayed streaming's mainstream adoption by several years. Yet paradoxically, the tragedy also underscored the internet's essential role in modern communication, setting the stage for its eventual resurgence. When YouTube launched in 2005 and Netflix began streaming in 2007, they entered a landscape that had been fundamentally reshaped by these twin crises.

For myself at Timberwolf Press and countless other media startups of that era, 9/11 marked a point of no return. The audiobook market would eventually recover and flourish, but the timing was devastating for early innovators who had invested ahead of the curve. Many of the formats and distribution methods we pioneered would later become industry standards – but through platforms and companies that emerged after the storm had passed.

What came next

That's where the next phase of streaming began. Not with broadcasters or corporations, but with everyday people.

YouTube certainly wasn't the first platform to offer video streaming, but it was the first to truly democratize it. For the first time, virtually anyone could have their voice heard without the need for radio station contracts, licensing negotiations, or broadcasting gatekeepers. All you needed was a camera, an internet connection, and a desire to share your story. YouTube wasn't just another website – it represented the next logical step forward from the foundations we laid during the Broadcast. com era. It opened the door to a new world where creativity, storytelling, and communication were no longer bound by traditional barriers. This paradigm shift didn't just change how we viewed entertainment; it fundamentally reshaped the internet and digital media forever.

In the sections that follow, we'll explore how the post-Broadcast. com era unfolded and how YouTube, Netflix, and Twitch took the streaming revolution to the next level.

Selling the Future: Inside Our Investor Roadshow

It's one thing to build a company. It's another to convince people to bet on it.

By the time we hit the investor roadshow, we knew we had something big. We'd solved the technical challenges, built a platform that worked, and proven there was an audience hungry for what we were offering. But investors? They weren't just looking for a great idea. They needed numbers, a growth story, and a clear path to making a return.

So, we took the show on the road. Boardrooms, hotel conference rooms, pitch meetings over awful coffee, we laid it all out. The market potential, the adoption curve, and the revenue model. Some investors got it immediately. Others took convincing. And some? Well, let's just say they missed out.

Here's what the roadshow taught us about selling a vision, about the investors who saw the future (and those who didn't), and about what it takes to turn an idea into a billion-dollar reality.

According to our **1999 roadshow deck** (I went and found it in my files), by the time Yahoo acquired us, we had:

- 385+ radio stations streaming live
- **40 TV networks** broadcasting over the web
- 420+ college and pro teams streaming live games
- 16,000+ live events annually
- **2,100+ business webcasts**, including earnings calls and keynotes
- **2,400+** Full-Length CDs
- 350 Live Concerts
- 65,000 Hours of On-Demand Audio and Video Programming

Our channels included: Auctions, AudioBooks, Business, CD JukeBox, Education, Entertainment, Live Radio, Live TV, Music, News, Public Affairs, Shopping, Special Interest, Spiritual, Sports, Technology, Video.

Our broadcast partners included: A. H. Belo Corporation, BBC World Service, Capitol Records, Capstar / SFX Communications, Clear Channel Communications, Granite Broadcasting, Host Communications, Learfield Communications, Major League Baseball, National Football League, National Hockey League, Turner Broadcasting: CNN/TCM/WCW, 150 Division IA Schools.

Today, streaming live events feels commonplace, even routine. But back then, every single broadcast was revolutionary. AudioNet didn't just pioneer internet broadcasting; we fundamentally changed what was possible. For the first time, fans didn't need a television or radio to experience their favorite sports live. Investors weren't tethered to phones or physical meetings for earnings calls, and businesses suddenly had the unprecedented capability to share corporate events directly to audiences worldwide – all streamed instantly and all accessible from a personal computer.

In the world of sports alone, AudioNet shattered conventions and set new standards. Just to highlight a few of these groundbreaking moments, we were the first ever to broadcast live over the Internet: First College Football: In September 1995, we broadcast the Southern Methodist University Mustangs vs. Arkansas Razorbacks game from the Cotton Bowl in Dallas, First College Bowl Games, MLB World Series, MLB All Stars, NFL Superbowl, NCAA Final Four, World Championship Wrestling, NY Yankee's Ticker Tape Victory Parade, The Masters / US Open Golf, Winter Olympics, World Championship Boxing, NHL Season & Stanley Cup.

In January '97, Super Bowl XXXI attracted the then-largest internet audience ever, with a webcast record of over half a million listeners.

In 1997, AudioNet hosted 50% of the live & on-demand streaming media on the internet. In '98, we passed a million viewers of Bill Clinton's testimony. By '99, we acquired SimpleNet and NetRoadshow. We launched broadcast.com Japan, where I was on the Board of Directors with Softbank. I recall a board meeting where Masayoshi Son, the Chairman and CEO of SoftBank, was fondly demonstrating an Aibo robot dog. You had to "adopt" the robot dogs, which I did for my boys shortly thereafter. □

Broadcast.com: Pioneering Community Building in the Streaming Era

Broadcast.com emerged as a trailblazer in online media, not only by delivering diverse audio and video content but also by fostering vibrant communities around shared interests. This approach marked a significant evolution from the earlier era of Bulletin Board Systems (BBS), which primarily facilitated text-based communication and file sharing.

From BBS to Broadcast.com: A Shift in Community Engagement

Bulletin Board Systems were prevalent in the 1980s, and early 1990s were platforms where users connected via dial-up modems to exchange messages, share files, and engage in text-based games. I recall dabbling in many of those with my dial-up modem. Anyone remember Delphi?

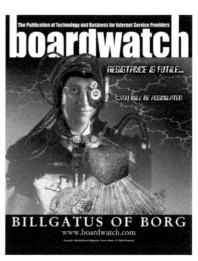

While they served as early hubs for niche communities, their capabilities were limited to text and static files, and their reach was confined to users with specific technical setups. I actually have a framed poster from the magazine of the day, Boardwarch, picturing Bill Gates as "Bill Gatus of Borg. https://en.wikipedia.org/wiki/Bulletin_board_system

In contrast, Broadcast.com leveraged the burgeoning internet infrastructure to stream live audio and video content, effectively bringing real-time events and media to users' screens. This innovation allowed fans to connect with their favorite sports teams, music artists, and other interest groups through live broadcasts, creating a more immersive and interactive experience. By offering live streams of events and shows, Broadcast.com enabled users to participate in real-time, fostering a sense of belonging and community. https://en.wikipedia.org/wiki/Broadcast.com

Building Communities Through Streaming

Broadcast.com's strategy went beyond merely providing content; it was about connecting people to their tribes. Whether that meant fans connecting with sports teams, niche groups bonding over specific content, or even the early days of social media communities, Broadcast. com built on the idea that streaming connects people to their interests and their peers. This approach laid the groundwork for future platforms like **YouTube**, **Instagram**, and **Twitch**, which have since become central hubs for community engagement through streaming.

Impact on Modern Streaming Platforms

The community-centric model pioneered by Broadcast.com has been instrumental in shaping the evolution of modern streaming platforms. Services like YouTube and Twitch have expanded upon this foundation, offering interactive features such as live chats, user-generated content, and community-driven events, thereby enhancing user engagement and fostering vibrant online communities.

In summary, Broadcast.com's innovative approach to streaming not only transformed media consumption but also set a precedent for community building in the digital age. By connecting individuals through shared interests and real-time content, we paved the way for the dynamic, community-driven platforms that define the streaming landscape today.

AudioNet.com showed the way with soundcards and encoders, but Broadcast.com connected people to their tribe, be it a football team or a Victoria's Secret Model.

The First Pre-Roll Streaming Ad: A Creative Solution to Content Theft

In the early days of AudioNet, we faced a significant challenge. Other websites were deep-linking our streams and effectively stealing our content without proper attribution or permission. At the time, there was no way to prevent this from happening, and we were losing valuable traffic and control over our brand.

Frustrated by this situation, I decided to get creative. I went out to my car and used my Pearlcorder (hand-held micro tape recorder – remember, no smartphones back then) to record a 7-second clip that simply said, "Thank you for listening to AudioNet. THE Broadcast Network on the Internet, at www.audionet.com." We began playing it before all of our streams. I still have the original "thanks.ra" audio file. The conversion is staticky, but it certainly brings back memories. https://www.patrickseaman.com/thanks-ra/

This brief message served a dual purpose: it shamed those who were stealing our content while also reminding users where the original stream was coming from. Little did we know (as far as I know) that this would be the first pre-roll ad ever to run in streaming media.

Interestingly, the response was immediate and overwhelming. While this was happening, the traffic to AudioNet surged, and our brand recognition grew. However, as our streaming audience continued to expand rapidly, we realized that those 7 seconds of airtime were costing more than we had anticipated, as those seven seconds used bandwidth every time it played. And bandwidth was very expensive.

Eventually, we moved from this "homemade" solution to running "real" ads, but that initial clip marked a milestone in advertising in the streaming era. It was the beginning of an entirely new way to think about monetizing content and creating a direct connection with the audience. By 1999, we were doing 4800+ spots per week on our radio station streams alone.

Robert Davidman sold the first preroll ad to Qualcomm. The second one was to Microsoft.

"I remember when we came up with the idea. I pitched the concept that "What if we can guarantee that before someone hears the radio station, that they must listen to your ad first" This concept was mind-blowing to advertisers. Then we used SMIL to make the client's website open up while the ad played." – Robert Davidman.

This moment was more than just about stopping theft. It marked a monumental milestone in the advertising ecosystem that would eventually become a critical part of how streaming platforms like YouTube, Hulu, and Spotify would evolve.

Side Story: Pioneering Audiobooks and the Birth of the Podcast: Timberwolf Press and the Internet Revolution

At AudioNet, we were always desperate to add more and new content. In 1997, I had an idea that would combine two passions of mine: storytelling and technology. I convinced my friend and fellow writer, the late Jim Cline, to produce an audiobook version of his epic science fiction novel, "A Small Percentage." Jim was a talented writer, and the thought of bringing his work to life in a new format excited both of us.

Jim built a small recording studio in his house, using friends and family to help with the production. Working on the project in our spare time, he recorded the story in weekly 30-minute episodes, a format inspired by the old radio serials that had once captivated listeners. What we didn't know at the time was that we were essentially inventing the podcast. This method of releasing episodic content in a serialized format was virtually unheard of in 1997, but now, it's a staple of modern media.

We were, as far as I know, the first to release an audiobook on the internet. Every week, we built an enthusiastic audience that was eager for the next episode. However, some of the Texan-sounding aliens we voiced in the series became a running joke, with listeners bitterly complaining about the voice acting in one breath and clamoring for the next installment in the next. The audience's engagement was incredible, and it was clear that the model we were using had struck a chord.

By the time the series was complete, it comprised 42 half-hour episodes. We later went so far as to commission storyboard art that was timed to present at designated timecodes in the playback.

After Yahoo's acquisition of Broadcast.com and the subsequent shift in the media landscape, Jim and I decided to take what we'd learned and apply it to a new venture. We co-founded Timberwolf Press, a new media company focused on creating full-cast dramatized audiobooks. This latest venture allowed us to push the boundaries of storytelling even further, and our work was recognized with multiple awards for its innovative approach to audio entertainment.

Not only did we create full-cast audiobooks, but (again, so far as I know) we were also the first to release them on MP3 CD, a format that allowed listeners to enjoy high-quality audio in a more accessible and portable way. Each of these releases also included ebooks on the CD-ROM, further enhancing the experience and embracing the potential of digital media. Our audiobooks weren't just content; they were experiences designed for the emerging digital landscape.

What Jim Cline and I created at Timberwolf Press in the early 2000s, with our serialized audio stories and full-cast dramatizations, essentially laid the groundwork for what would later evolve into the modern podcast industry. When we produced "A Small Percentage" as weekly 30-minute episodes with multiple voice actors, sound effects, and music, we were inadvertently creating what would today be called a "scripted fiction podcast" – a format that has exploded in popularity with shows like "Welcome to Night Vale," "The Black Tapes," and productions from networks like Gimlet Media and QCode.

Our distribution method – releasing episodic content directly to listeners over the internet – anticipated the subscription-based model that podcasts would later adopt through RSS feeds. We were solving the same fundamental challenges that podcast creators face today: how to serialize content, maintain consistent audio quality, build a loyal audience, and deliver a compelling narrative experience without visual elements.

Similarly, our approach to audiobooks presaged much of what

makes services like Audible successful today. When traditional audiobook publishers were still thinking about physical media (primarily cassettes and CDs), we were focused on digital distribution, enhanced production values, and the unique advantages of the audio format. Our decision to use full casts rather than single narrators was particularly forward-thinking – a premium approach that Audible would later adopt for its Audible Originals and high-profile titles.

The MP3 CD format we pioneered, combining audiobooks with ebooks on a single disc, conceptually anticipated how digital content is bundled today. When Amazon launched Whispersync for Voice in 2012, allowing users to switch seamlessly between reading and listening to a book, they were implementing a more sophisticated version of what we had attempted with our hybrid releases a decade earlier.

What's particularly striking is how the economics have evolved. When Timberwolf Press was active, producing high-quality audio content required significant upfront investment in studio time, voice talent, and engineering expertise. Today, the democratization of audio production tools has dramatically lowered these barriers to entry, enabling the explosion of podcasting we've witnessed since 2014. The same technological evolution that made our early work challenging has now created an environment where independent creators can reach audiences that would have been unimaginable during the Broadcast. com and Timberwolf Press era.

In many ways, our early experiments with audio storytelling and digital distribution were the bridge between old-world media and today's creator economy – showing what was possible when traditional narrative forms embraced the potential of digital delivery.

At the time, audiobooks were sold in large packages of cassette tapes or Audio CDs – remember, CDs could only hold about 74 minutes of audio, so you needed a bunch of them to hold a whole audiobook. We packaged the whole audiobook & ebook on a CD-ROM and sold them in DVD-style boxes at Barnes & Noble and other retailers. We also offered our audio in downloadable files that people enjoyed on their portable MP3 players. Remember those?

Revolutionizing Music Streaming: The Impact of RealAudio Codecs

In the early days of streaming, audio codecs were primarily designed for voice transmission, resulting in poor sound quality for music. RealNetworks released the RealAudio 1.0 codec in April 1995, which was capable of streaming voice-quality audio over the internet. The introduction of RealAudio's G2 codec in 1997, also known as the Cook codec, marked a significant advancement, enabling the streaming of music with acceptable fidelity.

Seizing this opportunity, I brought a CD jukebox from home

Elm Street T1: Machines
(199.179.200.xxx)

Non-PNServers
- .50 ww2.audionet.com / Netscape Web Server
- .58,100,101,102,103,104 / .58 Netscape Web Server
- .55 email server / mailhost.audionet.com:2541
- .53 CHAT server
- .88 UNIX DNS Server
- .54 SQL Server
- .88 Patrick's PC
- .85 Keith Brown
- .93 Richard Pak

Radio Stations
- .80 KZPS Max=100
- .61 KAAM Max=100
- .64 KPLX Max=100
- .56 KLIF Archive Max=100
- .81 WTEM Max=100
- .82 WMVP Max=100
- .76 KOME Max=100
- .71 KDGE Max=100

Other PNServers
- .68 Jukebox Max= 30

(remember those players that had a horizontal tray that you could load 5 CDs on and connect it to your home stereo system and connected it to a live encoder. I loaded it with several music CDs and began streaming live. The response was overwhelming; our servers were inundated with listeners eager to experience music streaming firsthand.

Let me rephrase that.... **Our traffic EXPLODED!**

Honestly, I was kind of shocked at the sheer SCALE of the reaction.

This experiment not only demonstrated the public's appetite for streaming music but also highlighted the potential of the Internet as a platform for music distribution. It was a pivotal moment that showcased the feasibility of streaming music to a broad audience, laying the groundwork for future music streaming platforms.

Fun fact: Do you know who made up a huge part of AudioNet/  Broadcast.com's daytime audience in the late 1990s? Office workers – lots of them, sitting in front of their desktop computers, discreetly tuning into our streams on headphones. Back then, radios were often prohibited by corporate policies, and even where permitted, radio reception was notoriously unreliable deep inside office buildings. As a result, our internet broadcasts quietly became the go-to solution for thousands of people who wanted to follow their favorite sports teams, news updates, or events during working hours. Of course, most of them weren't exactly advertising their listening habits to management, making AudioNet/Broadcast.com a bit of an open secret in offices around the world.

The success of this initiative underscored the importance of technological innovation in content delivery. Advancements in audio codecs, such as RealAudio, were instrumental in making streaming a viable and attractive option for both content creators and consumers. This development paved the way for the proliferation of streaming platforms that would revolutionize the music industry in the years to come.

Closing Thoughts: The Road Ahead for Streaming

From the dial-up days of Broadcast.com to YouTube, Twitch, and Netflix, streaming has come a long way. What started as a scrappy experiment, wiring up radio stations, encoding files by hand, and rebooting servers in the middle of the night, became the foundation for the media revolution we see today.

But if history has taught us anything, it's that no platform, no company, no technology stays dominant forever. In the same way that YouTube and Netflix changed the game, the next shift is already underway. AI-driven content, decentralized platforms, and, interestingly, real-time community engagement are all redefining what streaming can be.

The core truth remains: People want access. They want control over what they watch, when they watch, and how they interact. The platforms that adapt to that reality will thrive. The ones that don't? Well… just ask Yahoo! how that played out.

What's Next?

I was fortunate enough to be there at the very beginning of the streaming revolution – but this isn't merely a story about the past; it's a dialogue about what's coming next. Streaming has transformed not just our technology but our culture and expectations, and now we find ourselves at another crossroads. Have we reached the point of peak platform saturation, where innovation slows, and consolidation becomes inevitable? Or is the next groundbreaking platform – the next YouTube, Netflix, or Twitch – already quietly forming inside some ambitious startup's garage?

This brings me to the subject of the next chapter, where I'll explore one of the most profound shifts that streaming has catalyzed: our move from ownership to access. It wasn't long ago that we filled shelves with DVDs, CDs, vinyl albums, and tapes. Today, shelves have been replaced by subscription services, monthly payments, and cloud storage. Streaming hasn't simply reshaped how we consume content; it has fundamentally altered our relationship with media itself.

The shift from ownership to access didn't happen overnight; it unfolded gradually over decades. I saw this evolution firsthand while building Broadcast.com and even earlier while co-authoring the book "Website Sound" for New Riders/McMillan in 1996 alongside Jim Cline. We poured countless midnight-to-dawn hours into that project, capturing ideas that would soon become central to the streaming explosion that followed. It was an exhilarating time – juggling writing sessions between midnight and 6 AM while riding the thrilling wave that was AudioNet.

Yet, amid all the convenience and immediacy that streaming provides, we might pause to ask: Have we lost something valuable in this trade-off? Has the permanence and tangibility of ownership been sacrificed too easily in the name of convenience?

So, I leave you with a question to ponder: Do you still own your media in the traditional sense, or have you fully embraced the streaming economy and its promise of instant, endless access?

2.5: Interlude: Blockbuster, Enron, and the Delayed Future of Streaming

Prologue: The Stage is Set – A Cast of Giants

Before the first frame of Netflix buffered across a home internet connection, a different set of players had already stepped onto the stage. They weren't startups or disruptors. They were titans, wealthy, dominant, and in some cases, arrogant. But each would help define what streaming media could become, or more importantly, what it would have to survive to get there.

This is their story, not as protagonists or antagonists, but as early actors in a drama that would shape the digital media age.

Act I: Blockbuster – The Titan of Friday Night

The first to enter, with commanding presence and broad commercial appeal, was Blockbuster. Founded in 1985 in Dallas, Texas, Blockbuster grew from a single video rental store into a global juggernaut. With over 9,000 stores at its peak, it became the centerpiece of American weekend rituals.

Blockbuster didn't just dominate, it reshaped the home entertainment economy. It systematized video rentals, nationalized a fragmented industry, and generated billions in annual revenue. Its iconic blue-and-yellow branding and "Be Kind, Rewind" culture were embedded in the American psyche.

By the late 1990s, Blockbuster had everything a digital transition required: brand loyalty, physical inventory, subscriber data, and cash. But it lacked one thing, an appetite for self-disruption. Even as it dabbled in digital projects and overseas expansion, it remained anchored to its brick-and-mortar model. Its eventual partnership with Enron would be one of several bold missteps in its halting journey toward a digital future.

During my AudioNet/Broadcast.com years, I couldn't help but marvel that two of the largest media entities in the world had both emerged here in Dallas, Texas. One started in a spare bedroom at Mark's house, the other commanded thousands of retail locations.

For a brief moment, Dallas stood at the intersection of analog legacy and digital ambition, with Blockbuster on one side and Broadcast.com on the other. Though their paths soon diverged, with one acquired by Yahoo and the other failing to adapt to the coming wave of streaming, their shared origin, far from Silicon Valley or Manhattan's media corridors, highlighted Dallas's fleeting but pivotal role in the opening act of the digital media revolution.

Act II: Enron – The Smartest Guys in the Boom

Then came Enron, a swaggering Houston-based energy firm with ambitions far beyond its oil and gas roots. Through the 1990s, Enron reinvented itself as a trading platform for everything from electricity to broadband. Its executives, led by Jeffrey Skilling and Kenneth Lay, championed the idea that markets, not molecules, were its true product.

By 2000, Enron was the seventh-largest company in the United States by revenue, valued at more than $70 billion. Fortune magazine named it "America's Most Innovative Company" six years in a row. But innovation, in Enron's case, often meant invention, in both the visionary and fictional senses.

With the launch of **Enron Broadband Services (EBS)** in 1999, the company declared it would revolutionize the internet just as it had energy markets. Enron planned to stream full-length films, secure real-time data, and build a nationwide broadband trading platform. To investors and partners like Blockbuster, this seemed not just possible, but inevitable.

But there was a problem. Enron's broadband empire was largely smoke and mirrors. The infrastructure was incomplete, the technology untested, and the financials massaged beyond recognition. Still, in the feverish optimism of the dot-com era, few looked too closely. When Enron collapsed in 2001, it wasn't just a scandal, it was a catastrophic breach of trust in tech-enabled ventures.

Act III: Broadcast.com and Yahoo – The Signal Before the Storm

In a separate wing of the theater, another set of players had already delivered their lines. Broadcast.com, the company I helped build from the ground up as its first employee, had already proven that streaming live audio and video over the internet wasn't just possible. It was valuable, scalable, and in high demand. Its 1999 acquisition by Yahoo for $5.7 billion signaled Wall Street's belief that digital distribution was not just coming, but here.

What's often overlooked is that Yahoo was briefly positioned to be the digital distributor of Blockbuster content, nearly seven years before Netflix would stream its first film. Around July 2000, discussions emerged between Yahoo, Blockbuster, and Enron Broadband Services to create an exclusive partnership. Under this deal, Yahoo would be the online portal for ordering Blockbuster video rentals, with the potential for future streaming through Enron's fiber network.

David Cox, who reported to Jeff Skilling, led Enron's deal team. On the Blockbuster side was then-CEO John Antioco. I'm told that Enron structured the agreement so that costs would be rolled into a balloon note it held with Blockbuster, allowing Enron to book projected profits from the arrangement using its mark-to-market accounting model. Blockbuster, in turn, could claim immediate financial benefit on its books. The proposal promised minimal short-term cost to Yahoo, with Enron assuming the financial risk.

Technical and business development teams from Yahoo had already met with Enron earlier that spring in Dallas. A turning point came when an internal analyst at Yahoo drew a dry-erase diagram echoing what would later be called the 'Raptors,' a mechanism Enron used to generate paper profits by cycling assets through affiliated entities. The structure raised red flags for those paying attention.

On a technical level, Yahoo's networking staff struggled to validate the viability of Enron's proposed infrastructure. The CTO of Enron Broadband Services lacked visibility in the broader networking community, and there were questions about the readiness of their national backbone. A call with Yahoo's technical team left doubts about the network's buildout and capacity.

Eventually, an urgent call came in: the deal needed executive sponsorship. Several high-level meetings followed, including one in California that brought Cox, Antioco, and Yahoo's senior team to the table. But the deal collapsed over exclusivity terms. Yahoo demanded a significant financial commitment, and neither Blockbuster nor Enron could deliver it. There also lingering uncertainty about Blockbuster's streaming rights from the studios. While it was clear they could handle online rentals, streaming first-run titles was less certain.

Had the deal succeeded, it might have forced earlier legal and market reckoning over streaming rights, distribution control, and studio relationships. Even in failure, the attempt spurred conversations in Hollywood, which began bracing for what would become a decades-long disruption of traditional media distribution models. The idea that Blockbuster might lead the digital rental revolution, amplified by Yahoo's web presence and Enron's supposed infrastructure, sent early signals that the ground was shifting.

While it's impossible to say how much bandwidth would have actually been consumed or how viable the service would have been technically, the proposed partnership could have significantly accelerated debates around content licensing, distribution rights, and the technical feasibility of streaming to consumer households. Yahoo's platform and Broadcast.com's infrastructure were already capable of handling low-bitrate streams, and with proper integration, may have become one of the first major platforms for digital movie rentals. But timing, rights uncertainty, and Enron's collapsing credibility left the deal dead before launch.

Act IV: Setting the Stage – A Promise of Disruption

In the late 1990s, the concept of delivering movies directly to consumers over the internet was gaining traction. Broadcast.com's acquisition by Yahoo symbolized Wall Street's belief in digital media. Concurrently, Enron and Blockbuster saw opportunities in this emerging space, but their execution would have lasting consequences beyond their own downfalls.

Act V: The Enron Mirage – Streaming via Vaporware

Enron Broadband Services (EBS) aimed to revolutionize content delivery by creating a high-speed fiber-optic network. In 2000, Enron and Blockbuster announced a 20-year agreement to provide video-on-demand services over this network. However, the technology was not ready for mass deployment, and the partnership struggled to secure sufficient content licensing.

When the deal collapsed in March 2001, Enron's stock price suffered, and the company resorted to unethical accounting practices to mask the losses. This contributed to the broader Enron scandal, leading to one of the largest bankruptcies in U.S. history.

Act VI: Blockbuster's Blind Spot – Or was it?

Blockbuster is often remembered as a company that failed to adapt, frozen in its brick-and-mortar glory while the digital age surged ahead. That narrative holds some truth, but it oversimplifies the story. In reality, Blockbuster did attempt bold moves toward digital transformation, perhaps none bolder than its 2000 agreement with Enron.

The 20-year deal between Blockbuster and Enron Broadband Services was announced with fanfare. Together, they would stream movies directly to consumers' homes, bypassing physical media entirely. For Blockbuster, it was a genuine attempt to leap into the future. For Enron, it was an opportunity to showcase a bandwidth delivery platform it claimed would rival anything on the market.

But the foundation was flawed from the start. Enron lacked the technical infrastructure to deliver on its promises. Its broadband network was incomplete, its streaming capabilities unproven, and its sales pitch powered more by financial engineering than operational substance. The system didn't scale, and the pilot markets quickly revealed severe limitations in quality and reliability.

Behind the scenes, Enron compounded the damage. The company booked over $110 million in projected future revenue from the Blockbuster deal and promptly used that "revenue" in a convoluted accounting scheme, bundling it into off-balance-sheet partnerships to inflate earnings. When Blockbuster quietly exited the deal just months later, the promised revenue vanished, but Enron had already sold that revenue stream to outside investors. The damage was done. The move became part of the broader web of deception that led to Enron's collapse, but it also left Blockbuster exposed.

The failure wasn't merely technical. It was reputational. Blockbuster had staked its first major digital bet on a partner that imploded in scandal. The collapse of the Enron deal set back Blockbuster's digital momentum at a critical time. The internal appetite for digital risk cooled.

Resources were redirected. And when the company had a second chance to engage with Netflix in 2000, it declined the opportunity.

In retrospect, Blockbuster didn't fail because it had no digital ambition. It failed, in part, because one of its most visible attempts to evolve was tied to a partner whose collapse dragged others down with it.

Act VI.5: A Second Digital Push – The MovieLink Gambit

While the Enron debacle left Blockbuster hesitant, it didn't mark the end of its digital ambitions. In 2007, under new CEO Jim Keyes, Blockbuster made a bold move to acquire **MovieLink**, a studio-backed video-on-demand platform. The acquisition occurred within Keyes' first six months, signaling an early and serious commitment to pivot toward streaming.

Unlike Netflix's subscription model, MovieLink's architecture focused on **pay-per-view new releases**, aligning with Blockbuster's historical strength: freshly released titles. Roughly 80 percent of the company's rental volume had always come from new movies. The plan was to anchor a digital strategy around those high-demand titles while later layering in long-tail catalog content.

Keyes envisioned a **hybrid model** in which new releases were monetized through transactional VOD, while older titles were offered free or via subscription. Blockbuster would serve as the digital **aggregator of all movies**, forging distribution partnerships with major platforms like Google, Apple, DirecTV, and Dish Network.

At one point, Blockbuster and Google reached a **definitive agreement**: YouTube would serve as the destination for free content, while Blockbuster would provide the paid catalog. But just before launch, Google pulled out, reacting to **false rumors of Blockbuster's bankruptcy**, which derailed a potentially transformative deal.

A second, even more ambitious deal with **Viacom** would have secured exclusive digital rights to over 60 percent of long-tail content in the industry. But amid the 2008 economic collapse, Blockbuster's board rejected the deal. Spending $100 million annually on content licensing at a time of collapsing credit markets and uncertain consumer bandwidth access was seen as too risky. While history might judge that decision harshly, in 2009, buffering and limited broadband access made the future of streaming commercially uncertain.

The final blow came after Blockbuster's sale to Dish Network. Charlie Ergen, Dish's founder, had ambitions to use the Blockbuster brand as the core of a streaming-centric telecom offering, leveraging accumulated wireless spectrum. Plans were drawn up to acquire Sprint or T-Mobile and launch a next-gen wireless service with Blockbuster as the anchor content brand. But delays in regulatory bandwidth release halted the strategy. Eventually, Dish shelved the initiative and the Blockbuster brand with it.

Far from being a passive casualty, Blockbuster under Keyes pursued serious streaming strategies. What ultimately defeated those efforts wasn't vision, but market timing, financial instability, and the infrastructural realities of the late 2000s.

Act VII: The Fallout – A Trust Deficit and the Quiet Period

The high-profile failures of Enron and Blockbuster cast a shadow over the streaming media industry. Investors became wary of funding new ventures in this space, fearing similar outcomes. This skepticism led to a period of stagnation, delaying the development and adoption of streaming technologies.

But the damage did not end there. Enron's collapse caused ripple effects far beyond its own balance sheet.

One major casualty was **Rhythms NetConnections**, a broadband internet startup in which Enron had invested heavily. To hedge that investment, Enron's CFO created off-balance-sheet entities that masked losses. It was a part of the broader financial manipulation that led to Enron's implosion and further eroded investor confidence in tech-adjacent startups.

The fallout also brought down **Arthur Andersen**, Enron's accounting firm. At the time, Arthur Andersen was one of the "Big Five" global accounting firms, alongside names like Deloitte and PricewaterhouseCoopers. Its collapse was almost unthinkable. Complicit in document destruction and accounting misrepresentation, the firm lost its license and was forced to dissolve. This shook the foundations of corporate auditing, leading to the Sarbanes-Oxley Act and a more skeptical investment climate for technology firms.

Enron's effort to **commoditize bandwidth,** creating a futures market for data capacity, also left wreckage in the telecom industry. The company's aggressive push distorted supply and demand signals, fueled overinvestment in fiber infrastructure, and contributed to a broader telecom crash. Their failure poisoned the well for future digital infrastructure ventures and stalled the idea of bandwidth exchanges for years.

---

*These cascading effects helped cement the
perception that broadband-enabled streaming was
either a pipe dream or a risky gamble. That perception
would take nearly a decade to shake.*

Act VIII: What Was Delayed

The collapse of the Enron-Blockbuster partnership did more than end a corporate experiment. It stalled the evolution of the streaming industry at a critical moment when curiosity was turning into demand, and early infrastructure was beginning to catch up with imagination.

- **Video-on-demand at scale:** In theory, the Enron-Blockbuster deal could have launched the first major nationwide video-on-demand service years before Netflix pivoted to streaming. Instead, its failure signaled to both investors and infrastructure providers that the technology wasn't ready. In truth, the problem wasn't bandwidth capacity. It was execution, partnerships, and trust (and malfeasance on a grand scale on the part of Enron). The early failure slowed investment in edge delivery networks, discouraged telecom companies from expanding fiber access, and put a chilling effect on ambitious VOD startups that might have followed.
- **Consumer acceptance:** For the average household in 2001, the promise of streaming a movie over the internet went from thrilling to suspect. News coverage of the failed

deal, Enron's collapse, and Blockbuster's retreat from digital reinforced the notion that streaming was unreliable or "too early." The skepticism wasn't just about technology. It was about feasibility. Consumers didn't want to be guinea pigs for services that couldn't deliver the same reliability they were used to from DVDs and cable.

- **Content licensing innovation:** Perhaps the most lasting delay came from Hollywood. Studio executives, already wary of piracy and erosion of theatrical revenue, took the Enron-Blockbuster failure as a warning sign. If even a powerhouse like Blockbuster, backed by a multibillion-dollar energy firm, couldn't make digital distribution work, why should they license their crown jewels to unproven platforms? The result was a years-long period of retrenchment. Studios doubled down on physical media and restrictive licensing windows, inadvertently creating the very market conditions that allowed Netflix and Apple to later leapfrog them with more aggressive, customer-first models.

The ripple effects were clear. What might have been a digital renaissance in the early 2000s became a quiet, cautious detour. It would take another half-decade and a new cast of players to get streaming back on track.

Act IX: What Finally Broke the Logjam

The path from failure to acceptance wasn't immediate. Despite the early promise of digital media and the collapse of companies that tried to deliver it too soon or too recklessly, it would take several more years and a new generation of platforms for streaming to finally gain traction.

Netflix was already in the picture by 2000, but it was still a DVD-by-mail business struggling to survive the dominance of Blockbuster. Co-founder Reed Hastings famously approached Blockbuster that year with an offer to sell the company for just $50 million. Blockbuster laughed him out of the room. At the time, that decision seemed entirely rational. Netflix was losing money. It relied on the U.S. Postal Service. And streaming, at that moment, was viewed as speculative and immature. Why buy a company built on a digital future that had just failed so publicly through the Enron debacle?

But behind the scenes, Netflix was quietly preparing for that future. Hastings had always seen streaming as the inevitable endgame. The company focused relentlessly on user experience, data-driven recommendations, and a subscription model that reduced friction. Unlike Enron and Blockbuster, Netflix didn't try to skip steps. It waited for the infrastructure to catch up.

Between 2002 and 2007, three things changed:

Broadband became mainstream. In 2000, less than 10 percent of U.S. households had broadband. By 2007, that number had crossed 50 percent. Faster connections removed one of the biggest barriers to streaming: buffering.

The rise of new platforms shifted consumer expectations. YouTube's launch in 2005, followed by explosive viral growth, proved that millions of people would watch digital video: on demand, at scale, and outside traditional distribution models. It trained a generation of users to consume content online, from a browser, instantly.

Consumer devices caught up. The proliferation of laptops, smartphones, and later, smart TVs, created a user base that was no longer tethered to cable boxes or DVD players. Streaming could now go wherever the user was.

In 2007, Netflix launched its streaming platform. Initially limited and clunky, it was bundled with the DVD subscription as a kind of value add. But the seeds were planted. The very model that had failed spectacularly with Enron and Blockbuster, delivering video over the internet, was now viable. The key difference was execution, trust, and timing. Blockbuster, under new leadership, did try again: with MovieLink, with Google, with Viacom, and even with Dish. These weren't small ideas. They were serious bets on a future still held hostage by buffering, uneven access, and financial crisis. Netflix's success wasn't just about foresight. It was about timing, by plan or by fortune, until the environment was finally ready.

As Netflix began gaining traction, investor skepticism began to thaw. By 2010, streaming wasn't a longshot anymore. It was the future. Studios scrambled to catch up. Competitors emerged. And the second era of digital distribution, the one that finally proved successful, commenced in full force.

Curtain Call: Enron and Blockbuster as Anti-Models

Enron and Blockbuster stand as powerful warnings from business history. These weren't minor companies but billion-dollar industry giants with substantial funding, widespread respect, and considerable influence. Yet both ultimately crumbled due to overreaching ambition, inflexible business approaches, and critically, an inability to adapt to their rapidly evolving markets.

Enron had vision, but no discipline, or, perhaps, no scruples. It promised a broadband-fueled revolution before the roads were even paved. Its streaming ambitions were never backed by infrastructure, only by accounting tricks and investor hype. Blockbuster, in contrast, had the scale, the brand, and the consumer trust. It even made several serious attempts to reinvent itself digitally. But its boldest efforts came late, at a moment when financial headwinds and infrastructure limitations made execution impossible. In the end, the company was boxed in by a timing mismatch between its ambitions and the market's readiness.

What unites them is not just failure, but the type of failure: one born from overconfidence in paper empires, the other from underestimating a technological shift happening right under its nose. Both remind us that in the world of innovation, timing is everything, and that even the strongest balance sheet can't withstand a failure to adapt.

I watched it all unfold in real time: Blockbuster's hesitation, Enron's bravado, Yahoo's blunders. Each of them taught me something, not just about streaming, but about risk, momentum, and how fragile even the biggest giants can be when vision outpaces execution. These weren't just corporate collapses. They were signal flares, illuminating what happens when companies confuse size with permanence, or vision with inevitability.

In many ways, their failure cleared the stage. It gave space to the next generation of companies that would eventually build on the foundation we built at Broadcast.com. It would give broadband more time to expand. And, it would give investors more time to recover and rebuild confidence in digital media.

The curtain fell on Enron and Blockbuster. But the lessons they left behind still echo through every boardroom that dares to take the stage.

3: The Legacy of the On-Demand Revolution

Streaming and the Shift in Media Consumption: From Ownership to Access

The advent of streaming has fundamentally altered how people consume media, shifting from ownership of physical media to on-demand access. In the pre-streaming era, consumers relied heavily on physical formats like tapes, CDs, and DVDs to access music, movies, and television shows. However, the rapid growth of streaming platforms, beginning with pioneers like Broadcast.com and evolving through services such as Netflix and Spotify, disrupted this model.

The Decline of Physical Media

As streaming became mainstream, physical media sales plummeted. Between 2006 and 2019, U.S. DVD sales plummeted by 86%. According to the Recording Industry Association of America (RIAA), revenues from digital music streaming surpassed physical media sales for the first time in 2015, signaling a significant cultural shift.

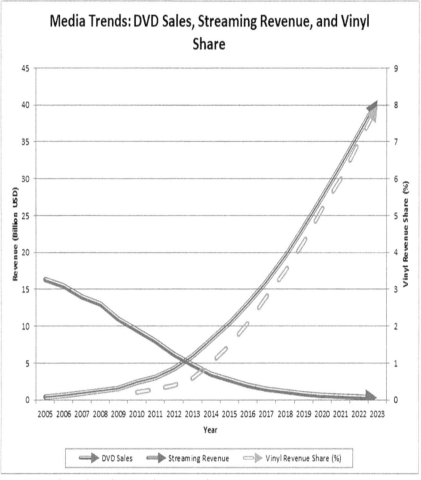

5: Graph: Decline of DVD Sales vs. Rise of Streaming Revenues (2005-2023)

- https://theweek.com/news/media/961866/death-of-dvds-and-decline-of-ownership-in-digital-age
- https://www.engadget.com/entertainment/streaming/vinyl-record-sales-continue-to-rise-amid-music-streamings-dominance-142131328.html
- https://en.wikipedia.org/wiki/Vinyl_revival
- https://www.statista.com/chart/7699/lp-sales-in-the-united-states/

Similarly, in the first half of 2023, physical media sales in the U.S. decreased to $754 million, down from $1.05 billion during the same period in 2022. This downward trend reflects a broader cultural shift towards digital consumption.

- https://theweek.com/news/media/961866/death-of-dvds-and-decline-of-ownership-in-digital-age
- https://www.riaa.com/reports/
- https://www.statsignificant.com/p/the-rise-fall-and-slight-rise-of

The Rise of Binge-Watching Culture

Streaming also introduced a new behavior: binge-watching. Platforms like Netflix revolutionized content delivery by releasing entire seasons of shows at once, encouraging viewers to consume hours of programming in a single sitting. A 2022 Nielsen report revealed that over 80% of global consumers engage in binge-watching, redefining the traditional episodic release model). This behavioral shift not only reshaped how content is produced but also set new expectations for viewer engagement. https://www.nielsen.com

Streaming and Consumer Freedom

Streaming has democratized access to content, allowing consumers to personalize their viewing and listening experiences. Algorithms tailor recommendations, creating unique journeys for each user. This level of personalization, paired with affordable subscription models, has made media more accessible than ever before.

The Vanishing Archive: Ownership in the Age of Digital Curation

In an era dominated by streaming services and cloud-based content, the concept of owning physical media has become an artifact of the past, reserved largely for collectors and preservationists. The shift from tangible formats, VHS tapes, DVDs, Blu-rays, and even digital downloads to ephemeral streaming libraries represents more than just a change in consumer habits; it is a fundamental transformation in how culture is curated, controlled, and consumed.

For many, the convenience of on-demand access outweighs the desire to own a permanent copy of a film, book, or album. Yet, this convenience comes at a cost: the impermanence of digital media. Unlike a physical book or disc that remains unchanged once produced, digital media exists in a fluid state, subject to updates, alterations, and even outright removal at the discretion of rights holders. What exists today in a streaming library might not be the same tomorrow or might disappear entirely.

This raises a question: If we do not own our media, who does? And how does this ownership influence the content itself?

Over the past few decades, we've seen multiple instances where classic works of film and television have been altered, sometimes decades after their original release. These changes, often made for reasons of political correctness, modern sensibilities, or directorial revisionism, can dramatically shift the way a work is perceived by future audiences.

Consider two of the most infamous cases in film history:

Star Wars: A New Hope (1977) – The original theatrical version featured Han Solo shooting Greedo first in the cantina scene, reinforcing Han's status as a morally ambiguous rogue. However, in 1997, George Lucas altered the scene so that Greedo shoots first (and misses), changing Han's actions to appear more defensive. This revision sparked significant backlash, as fans saw it as an unnecessary dilution of Han's character arc. Today, finding an official, high-quality version of the original theatrical cut is nearly impossible without resorting to bootlegs or fan restorations.

E.T. the Extra-Terrestrial (1982) – The 20th Anniversary re-release in 2002 digitally replaced FBI agents' firearms with walkie-talkies to make the scene less threatening for children. However, after widespread criticism, director Steven Spielberg later admitted regret and restored the original version in subsequent releases. While, in this case, the original was eventually reinstated, it highlights how artistic choices can be rewritten based on changing cultural attitudes.

These examples illustrate why some collectors and preservationists insist on owning physical media: they want to ensure access to the original versions, unaltered by retroactive changes. Unlike a streaming license, which can be revoked or modified, a physical copy remains a fixed record of its time, a snapshot of cultural history as it was rather than how later generations might wish it to be.

Modern Streaming Modifications: When Content Vanishes or Changes

The practice of altering or removing content hasn't diminished in the streaming era – if anything, it has accelerated. In 2020, HBO Max temporarily removed "Gone with the Wind" from its platform, later restoring it with a disclaimer about its historical context and racist depictions. That same year, Netflix pulled episodes of "Community," "The Office," and "30 Rock" that contained blackface or other content it considered racially insensitive. Disney+ added content warnings to classics like "Dumbo" and "Peter Pan," while subtly editing scenes in films like "Splash" to conform to family-friendly standards. Paramount+ modified a "SpongeBob SquarePants" episode that referenced a "virus," deeming it insensitive during the COVID-19 pandemic. Even music hasn't been immune – Spotify quietly replaced the original version of Kanye West's "Gold Digger" with a 'clean' edit without notifying listeners. In each case, consumers who believed they had "access" to these works discovered that this access was conditional, subject to changing cultural standards, corporate decisions, or licensing agreements. Unlike physical media owners, streaming and other digital media subscribers, including for ebooks, have no recourse when content is altered or vanishes overnight – highlighting a fundamental vulnerability in the access-based model.

The decline of physical ownership also poses challenges beyond content alteration. Licensing agreements dictate what is available on streaming platforms, and films shows, or entire catalogs can vanish overnight due to corporate decisions, contract expirations, or controversies. Some films that were once widely available have become nearly impossible to find outside of private collections.

In this sense, the role of personal ownership extends beyond nostalgia. It becomes an act of cultural preservation. While media companies view ownership as obsolete in the age of subscriptions and ad-driven platforms, those who value the integrity of historical works increasingly see physical media as a bulwark against revisionism, censorship, and corporate control.

If the trend continues, the ability to access unaltered versions of media may rest not with the general public but with private collectors, niche archivists, and underground preservation groups. In the end, the question is not just whether we still need to own media but whether we can trust that the media we consume today will remain the same tomorrow.

The Vinyl Revival: A Niche Resurgence in the Digital Age

While the decline of physical media has been a dominant trend in recent decades, an intriguing counter-movement has emerged: the resurgence of vinyl records. This revival, though significant, remains a niche market within the broader landscape of music consumption.

In 2022, vinyl records experienced a 17% growth in revenue, reaching $1.2 billion. Their sixteenth consecutive year of growth. This marked the first time since 1987 that vinyl outsold CDs, accounting for nearly three-quarters of physical music revenue. https://www.npr.org/2023/03/10/1162568704/vinyl-outsells-cds-first-time-since-1987-records

This resurgence is driven by a combination of nostalgia and a desire for tangible music experiences. Notably, younger generations, including Gen Z, have contributed to vinyl's renewed popularity, often viewing records as collectible items. High-profile artists like Taylor Swift have also played a role; her album releases have significantly boosted vinyl sales. https://nypost.com/2024/12/03/lifestyle/vinyl-records-growing-more-popular-than-cds-streaming/

Despite these gains, vinyl remains a small segment of the overall music industry. In 2022, vinyl records accounted for 8% of recorded music revenues in the U.S., with streaming services dominating the market. https://www.statista.com/chart/7699/lp-sales-in-the-united-states/

The vinyl revival underscores a nuanced shift in consumer behavior. While digital access and streaming services have become the norm, a dedicated subset of listeners continues to value the tactile and auditory experience that vinyl provides. This trend highlights the enduring appeal of physical media, even as the industry moves increasingly toward digital formats.

Consumers Fight Back: Reclaiming Ownership in the Streaming Age

The vinyl revival is just one piece of a broader consumer pushback against the erosion of ownership in the digital media landscape. As streaming services like Netflix, Spotify, and Amazon Prime have taken over, offering vast libraries at the cost of permanence, a varied ecosystem of resistance, adaptation, and innovation has emerged among those determined to reclaim control over their media.

One clear trend is the growing interest in personal digital libraries stored on local devices. While exact figures are hard to pin down, the Recording Industry Association of America (RIAA) reported in its 2022 year-end review that physical formats, including vinyl and CDs, accounted for 11% of total music revenue, hinting at a sustained appetite for ownership even as streaming dominates with 84%. Beyond physical media, the popularity of Network Attached Storage (NAS) devices suggests a shift toward digital permanence. Synology, a leading NAS manufacturer, has noted increased demand for home storage solutions in recent years, though specific sales growth like a "28% increase since 2021" isn't publicly detailed in their reports. Still, their focus on media server capabilities – like support for Plex – points to a consumer base eager to manage their own digital collections.

6: NAS Storage Market Growth

- https://www.marketresearchfuture.com/reports/network-attached-storage-market/market-size
- https://www.zionmarketresearch.com/report/network-attached-storage-market
- https://www.skyquestt.com/report/network-attached-storage-market
- https://www.globenewswire.com/news-release/2024/08/16/2931566/0/en/Network-Attached-Storage-NAS-Market-Size-Worth-USD-109-72-Billion-by-2032-Driven-by-Rising-Data-Volumes-Increasing-Adoption-of-Hybrid-Storage-Research-by-SNS-Insider.html
- https://www.skyquestt.com/report/network-attached-storage-market

The open-source community has fueled this movement with

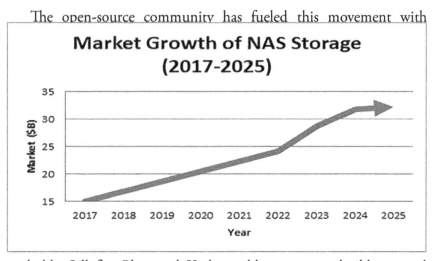

tools like Jellyfin, Plex, and Kodi, enabling users to build personal streaming services from locally owned files. Plex, for instance, has grown significantly, with the company announcing in 2021 that it had surpassed 15 million registered users – a number that likely climbed higher by 2022, reflecting strong demand for self-hosted alternatives. These platforms blend streaming's sleek interfaces with the reliability of ownership, letting users sidestep the whims of corporate content licensing.

Beyond DIY solutions, alternative business models are emerging to bridge the gap between access and ownership. Ultra HD Blu-ray sales, though a niche market, have demonstrated resilience amid the broader decline of physical media. According to the Digital Entertainment Group's (DEG) Year-End 2022 Digital Media Entertainment Report, total U.S. consumer spending on physical home entertainment (including DVD, Blu-ray, and 4K Ultra HD) reached $1.58 billion in 2022, down 19.77% from $1.97 billion in 2021. Within this shrinking market, 4K Ultra HD Blu-ray sales grew by 20%, driven by enthusiasts who value high-quality visuals and the permanence of physical ownership (FlatpanelsHD, February 9, 2023). Digital locker services like Movies Anywhere, launched in October 2017 by Disney and now supporting multiple studios such as Universal, Sony, and Warner Bros., allow users to consolidate purchased digital films across platforms. This offers a semblance of ownership, though it remains tethered to corporate ecosystems and subject to licensing agreements (Movies Anywhere, "About Us").

- Digital Entertainment Group (DEG) Year-End 2022 Digital Media Entertainment Report: Summarized in Media Play News (February 7, 2023) and FlatpanelsHD (February 9, 2023). Total physical spending reported as $1.58 billion, with 4K Ultra HD Blu-ray growth noted at 20%. Available via DEG press releases and secondary coverage (e.g., https://www.mediaplaynews.com, https://www.flatpanelshd.com) .
- FlatpanelsHD, "US digital home entertainment market grew 14% in 2022, DVD/BD down 20%" (February 9, 2023): Confirms DEG data, including the $1.58 billion figure and 4K Ultra HD growth.
- Movies Anywhere, "About Us": Official site details the service's launch in October 2017 and its studio partnerships https://moviesanywhere.com/about.

Content creators are also experimenting with direct-to-consumer approaches that emphasize permanence. Louis C.K. famously sold DRM-free downloads of his comedy specials starting in 2011, a model that let fans own their copies outright. Radiohead's 2007 In Rainbows pay-what-you-want release and Nine Inch Nails' 2008 Ghosts I-IV digital distribution similarly gave listeners high-quality, platform-independent files, proving there's an audience for unencumbered ownership.

Blockchain technology is reshaping music distribution through platforms like Audius, which leverages decentralized systems to connect artists and fans directly. Launched in September 2018, Audius allows artists to upload and share tracks without traditional intermediaries, giving them greater control over their work while exploring new ways to engage listeners, such as through its native AUDIO token and NFT integration (Coinbase, "How is Audius decentralizing the music industry?"). Blockchain-based media ownership, particularly via Non-Fungible Tokens (NFTs), has also gained attention, though it remains experimental and polarizing. According to DappRadar, NFT sales surged to $24.9 billion in 2021, a dramatic rise from $94.9 million in 2020, before plummeting to $3.4 billion in Q3 2022 amid a broader crypto downturn (Reuters, "NFT sales plunge in Q3, down by 60% from Q2"). This boom-and-bust cycle underscores a wider effort to redefine digital possession in the streaming era.

- Coinbase, "How is Audius decentralizing the music industry?" https://www.coinbase.com/learn/crypto-basics/how-is-audius-decentralizing-the-music-industry
- Reuters, "NFT sales plunge in Q3, down by 60% from Q2" (October 3, 2022) https://www.reuters.com/technology/nft-sales-plunge-q3-down-by-60-q2-2022-10-03/
- DappRadar, "Non-fungible Tokens: Sales Of NFTs Hit $25 Billion In 2021" (via Investors King, January 12, 2022) https://investorsking.com/2022/01/12/non-fungible-tokens-sales-of-nfts-hit-25-billion-in-2021/

Online communities further highlight this ownership drive. Subreddits like r/DataHoarder and r/PleX, with memberships of 500,000 and 600,000 respectively as of mid-2023 (per publicly visible counts), are bustling hubs for media preservationists. These groups swap tips on archiving everything from ripped DVDs to downloaded streams, often citing frustration with content vanishing from platforms – a trend backed by a 2023 Variety report noting over 200 titles removed from major streaming services in the past year alone.

Even casual users are adapting to the uncertainties of streaming. The 2022 Deloitte Digital Media Trends survey (16th edition) found that 36% of U.S. consumers who canceled a streaming video-on-demand (SVOD) service in the past six months cited "the content I want isn't available when I want it" as a key frustration, a sentiment tied to licensing shifts and content removal. This aligns with a broader behavior: many value offline access as a way to retain control over their favorites, especially as platforms like Netflix and Disney+ impose digital rights management (DRM) restrictions or expiration dates on downloads. This practice reflects a persistent desire to mitigate the impermanence of streaming libraries. Deloitte Digital Media Trends, 16th Edition.
https://www2.deloitte.com/us/en/insights/industry/technology/digital-media-trends-survey.html

These responses – from vinyl collectors to digital hoarders – show that streaming's convenience hasn't fully erased the appeal of ownership. Journalist Kyle Chayka, writing in The New Yorker in 2021, captured this tension: "Streaming promised infinite choice, but its impermanence has sparked a quiet rebellion for things we can hold onto." The push-pull between access and possession will likely shape media habits for years, as consumers carve out hybrid paths that marry streaming's ease with ownership's security.

The Birth of the On-Demand Generation: How Broadcast. com Pioneered Digital Media Consumption

Before streaming giants like Netflix and Spotify redefined media consumption, AudioNet/Broadcast.com was already laying the groundwork for a world without physical media dependency. In the mid-to-late 1990s, when most people still relied on CDs, VHS tapes, and terrestrial radio for entertainment, Broadcast.com was pioneering digital on-demand access, proving that consumers wanted more control over what they listened to and watched, even if they didn't know it yet.

Beyond Terrestrial Radio: The CD Jukebox and Vertical Programming

Unlike traditional radio stations that played a mix of content dictated by programming directors, Broadcast.com introduced a CD Jukebox model and the availability of more than 2,400 full-length CDs, allowing users to personalize content selection. This was a radical departure from the one-size-fits-all approach of terrestrial radio. Instead of tuning in to whatever was playing at the moment, listeners could choose their own audio experience, a precursor to today's streaming music platforms.

But Broadcast.com wasn't just about music. The company

experimented with on-demand audiobooks, live sports broadcasts, and video content long before these became mainstream digital services. At the time, Pay-Per-View (PPV) on cable was still in its infancy and mostly focused on special events and adult entertainment. Outside of niche applications, such as satellite services for businesses or expensive PPV events, on-demand content wasn't something most consumers engaged with.

Linear Programming vs. On-Demand: A Paradigm Shift

In the 1990s, nearly all media followed a linear programming model:

- Cable TV, OTA (Over-the-Air) Television, and Satellite followed pre-set schedules where viewers and listeners had to tune in at specific times.
- If you wanted a personal copy of a movie or music, you naturally had to own physical media, whether it was a VHS tape, a CD, or a cassette.
- If you wanted a book, you either had to own a physical paper book, the audiobook on cassette tape(s), or a set of audio CDs. eBooks were still embryonic. Anyone remember the Rocket E-Reader? When I published Jim Cline's epic *A Small Percentage*, we released it in hardcover, cassette tape, audio CD set, *and* MP3 CD-ROM that included the ebook alongside the audiobook. We also streamed the audiobook on AudioNet (the first-ever audiobook on the Internet) and later made it available as a downloadable audiobook and/or ebook.
- Pay-per-view (PPV) was available but limited in scope, expensive, and primarily used for one-time events.

Broadcast.com disrupted this paradigm by offering a glimpse of a future where consumers could access content whenever they wanted, without owning it physically. While the quality wasn't 4K or even HD by today's standards, the demand was there and growing. The ability to listen to a curated audiobook, select a sports broadcast, or choose a CD from a massive digital library was revolutionary.

Content is King – Then and Now

The transition from physical media to digital access was never just about technology. It was about **content**. Streaming, digital libraries, and on-demand services have changed how we access media, but they haven't changed the fundamental truth: **content is what drives audience engagement, not just the platform delivering it**.

"Content is King." It was on all our decks, and it was something Mark Cuban repeated every chance he got. (He later added, "Distribution is Queen.")

Back in 1996, years before streaming became the dominant force it is today, I wrote:

> *"No matter what technology you use, whether you stream or download, whether you are live or recorded, you are ultimately in a ratings war based on your content. Just ask the TV networks whether they would rather have the newest, most exotic technology or a guaranteed #1 primetime show."*
> *– Patrick Seaman, Website Sound (Circa 1996, New Riders/McMillan)*

This was true when television fought for dominance over radio, it was true when cable tried to dethrone network TV, and it remains true today in the battle between streaming services. Ownership of media may be fading, but the need for compelling content has never been stronger.

The most advanced technology in the world means nothing without programming that keeps audiences engaged.

At Broadcast.com, we understood this from the start. Our CD Jukebox model wasn't just about the infrastructure. It was about giving people choices they had never had before. We didn't just throw a collection of music online and hope for the best. We organized content into vertical genres, created specialized listening experiences, and treated digital media as something more than just a storage medium. It was an engagement platform.

Netflix, Spotify, and other streaming giants have mastered this philosophy. Their power comes not just from their libraries but from the way they curate and present content to keep audiences locked in. Whether it's the algorithm-driven recommendations of today or the genre-driven streaming experiments of Broadcast.com, the key to success has always been delivering the right content to the right audience at the right time.

But with digital curation comes another shift: who decides what content is accessible? As media ownership fades and access becomes the primary model, audiences are no longer the ultimate arbiters of what survives. This is where content control, alteration, and even revisionism come into play.

From Broadcast.com to the Streaming Era

The innovations at Broadcast.com didn't just challenge how content was consumed. They foreshadowed the rise of streaming media as a dominant force. The principles that Broadcast.com introduced, such as on-demand content, verticalized programming, and digital over physical media, became the foundation of services like Netflix, Spotify, and Audible.

Today, the very idea of ownership in media is fading in favor of streaming platforms that deliver music, films, books, and sports on demand. The shift from linear programming to consumer-controlled access started with early pioneers like Broadcast.com, proving that once people had the choice, they would never go back. The on-demand generation was born.

Building the Rocket: The Untold Complexity of Scaling On-Demand Media

Ingesting all that content wasn't just about flipping a switch and making magic happen. It took an ever-expanding array of hardware, infrastructure, people, and sheer willpower to keep up with demand. The further we pushed, the more we realized just how uncharted this territory was.

I vividly recall those early planning sessions, crowded around one of the back tables in our conference room (Café Brazil, down the

street) as we carefully mapped out what we believed would cover our infrastructure needs, servers, encoders, networking gear, and storage solutions. We'd diligently calculate our best estimates, and then, driven by caution and experience, we'd bump those projections even higher, thinking we'd accounted for every conceivable contingency.

Yet despite all this careful forecasting, our estimates consistently fell short. Reality quickly outpaced our most generous budgets because nobody had ever attempted something of this scale or complexity before or experienced the kind of exploding demand we saw every day. Each day brought unexpected demands, new challenges, and rapidly escalating costs that no one could have fully anticipated. The entire experience felt less like steering a controlled launch and more like strapping ourselves directly onto a rocket's boosters, lighting the fuse ourselves, and holding on as it surged skyward at full throttle.

Some aspects of the scale-up were obvious:

- The forest of at least 25 satellite dishes we installed on the roof at our 2914 Taylor Street headquarters to pull in live feeds.
- The ever-growing racks of servers and encoders stretched down long, cold aisles, humming with the heat of a thousand conversions happening in real-time.

However, the part that didn't get as much attention until it became a crisis was managing the physical media itself. Content owners sent us mountains of tapes, CDs, hard drives, and formats we sometimes had to reverse-engineer just to access. Every day, new crates of media arrived from sports leagues, audiobook publishers, music labels, and video producers, and there was no off-the-shelf system to track it all. No one had ever built a platform for managing digital ingest at this scale.

We had to invent everything from scratch:

- A tracking system for where each tape was, what stage of encoding it was in, and when it would be ready.
- A workflow that could handle multiple formats, varying quality levels, and a content pipeline that never stopped growing.
- A media storage and retrieval process that kept expanding at a rate that felt exponential.

We weren't just streaming audio and video. We were creating the infrastructure that made on-demand media even possible. It was our own moonshot, and every single day, we had to rethink, rebuild, and scale up just to keep from falling behind.

People talk about the birth of streaming as if it was inevitable. It wasn't. It was an unrelenting, white-knuckle fight to figure out what worked, what didn't, and how we could keep pushing forward before we ran out of bandwidth, storage, or sanity. We didn't call it a rocket ride for nothing.

Here's a section on how the shift from ownership to streaming has played out differently in global markets, providing an international perspective to complement the American-focused analysis:

Global Perspectives: How the Streaming Revolution Varies Worldwide

While the transition from ownership to streaming access has been a global phenomenon, its pace and impact have varied significantly across different markets, shaped by regional infrastructure, regulatory environments, and cultural attitudes toward media consumption.

In East Asian markets like South Korea and Japan, physical media has shown remarkable resilience alongside streaming growth. Japan still maintains a robust physical music market, with CD sales accounting for over 70% of music revenue as recently as 2022 – nearly ten times higher than the U.S. proportion. This persistence reflects both cultural collecting practices and artist-fan relationships where physical purchases are seen as direct support for creators. Similarly, South Korea's K-pop industry has strategically embraced physical albums as collector's items with photocards and exclusive content, even as streaming platforms like Melon and global services like Spotify gain traction.

European markets have taken different approaches to balancing streaming convenience with cultural preservation. France's "cultural exception" policies require streaming platforms to invest significantly in local content production and maintain quotas for European works. Meanwhile, Germany's strong tradition of public broadcasting has influenced its streaming landscape, with media libraries from public broadcasters competing alongside global platforms like Netflix and Amazon Prime.

Developing markets present a different picture altogether. In India, the leap to digital often bypassed the physical ownership phase entirely for many consumers. Millions of users experienced their first film or music streaming before ever owning a DVD or CD, creating a generation with no nostalgic attachment to physical media. Services like JioSaavn and Hotstar have adapted to local conditions with ultra-low-cost subscription tiers and significant free, ad-supported options that acknowledge economic realities while still building the access-based model.

Infrastructure limitations in regions like sub-Saharan Africa and parts of Southeast Asia have led to innovative hybrid models. Years after broadcast. com I experienced these challenges firsthand while working at Ole Media, where we faced significant hurdles implementing pay-per-view content in Mexico and the Philippines. The limited availability of credit cards in these markets meant that standard Western payment models simply weren't viable for most potential customers. We had to develop alternative payment systems using mobile credits, convenience store payment codes, and even cash-based voucher systems to make streaming content accessible. Similarly, at Cinsay, we encountered comparable obstacles throughout Eastern Europe, where banking infrastructure varied dramatically between urban and rural areas. These experiences taught me that the streaming revolution isn't just about technology – it's equally about adapting financial models to fit local economic realities.

In China, the streaming landscape evolved under unique regulatory conditions, with services like iQiyi, Youku, and Tencent Video dominating a market where global platforms like YouTube and Netflix remain largely unavailable. These services have pioneered alternative monetization strategies beyond subscriptions, including virtual gifting during live streams and integration with e-commerce platforms.

These global variations remind us that while the fundamental shift from ownership to access is universal, its expression reflects local economic conditions, cultural values, and technological infrastructures. As streaming continues to evolve, understanding these differences becomes increasingly important for both platforms seeking global expansion and consumers navigating an increasingly borderless media environment.

Streaming Economics: Creator Compensation in the Access Era

Beyond changing consumer relationships with media, the streaming revolution has fundamentally altered how creators, especially musicians, are compensated for their work. The shift from ownership to access hasn't just transformed how we consume content; it has completely restructured the economic foundation of creative industries.

In the ownership era, a musician earned a substantial one-time payment when a fan purchased their album, typically between $1-3 per CD sold. This model created predictable income streams where popular albums could generate significant revenue through sales alone. For example, an artist with a Gold record (500,000 units sold) could earn upwards of $1 million from album sales before considering concert revenue or merchandise.

A song must be played about 250 times to equal the artist's revenue from a single album sale...

Streaming has replaced this model with a system of micro-payments based on individual plays. As of 2025, Spotify pays artists approximately

$0.003-$0.005 per stream, meaning a song must be played about 250 times to equal the artist's revenue from a single album sale. Apple Music and Amazon Music offer slightly higher rates (around $0.007-$0.01 per stream), while YouTube's rates remain among the lowest at approximately $0.00069 per stream. These economics have created a "winner-takes-most" environment where truly massive streaming numbers are required for sustainable income.

The impact on artist earnings has been dramatic and uneven. According to a 2023 UK Parliament inquiry into music streaming economics, only the top 1% of artists can make a living wage through streaming alone. The report found that while global recorded music revenue has returned to near-CD-era peaks (reaching $25.9 billion in 2023), a significantly smaller percentage of that revenue reaches artists compared to the ownership era.

Mid-tier and niche artists have been particularly affected. Indie musician Zoe Keating publicly shared her streaming earnings, revealing that 2.6 million Spotify streams generated only $12,231 – a sum that would have represented only around 800 album sales in the pre-streaming world. For context, streaming 2.6 million times suggests a substantial fan base, yet the corresponding revenue barely covers a month's living expenses in many major cities.

This economic reality has forced artists to adapt. Tours and merchandise, once promotional tools to drive album sales, have become primary revenue sources for many musicians. Platforms like Bandcamp (which allows fans to directly purchase music in physical or digital formats) have grown as artists seek alternatives to streaming economics. Some have embraced Patreon or similar subscription services to establish direct financial relationships with fans. Even vinyl's resurgence can be partially understood as a response to streaming's economic model – artists earn substantially more from a vinyl sale than from hundreds of streams.

The economics of streaming aren't universally negative for creators. The model has virtually eliminated the barriers to global distribution, allowing independent artists to reach audiences worldwide without major label support. Streaming has also enabled some musicians to build sustainable careers through consistent, steady accumulation of streams across large catalogs rather than hit-driven sales spikes.

For filmmakers, writers, and other content creators, the streaming shift has created different economic challenges. Residual payments for actors and writers, once a reliable income source when shows were rerun, or DVDs sold, have diminished as streaming platforms pay fixed buyout fees rather than performance-based compensation. This tension was central to the 2023 WGA and SAG-AFTRA strikes, which sought to establish fair compensation models for the streaming era.

As we debate the merits of ownership versus access models for consumers, we must also consider how these models affect the creators whose work we enjoy. The convenience and breadth of streaming come with hidden costs – not just in terms of consumer control but in the changing economic foundation of creative industries themselves. Finding sustainable models that fairly compensate creators while maintaining the benefits of digital access remains one of the critical challenges of the streaming revolution.

Closing Summary for Chapter 3: The Legacy of the On-Demand Revolution

The shift from physical media ownership to digital access wasn't just a matter of convenience. It was a fundamental reshaping of how we consume, store, and value content. We didn't just wake up one day and decide that we didn't need CDs, DVDs, or hard drives filled with MP3s anymore. It was a slow but inevitable transition, one that companies like Broadcast.com helped accelerate by proving that on-demand access wasn't just possible. In many ways, it was better.

At the time, people didn't know they wanted streaming media because they had never had real access to it. Terrestrial radio and network television were still dictating schedules, Pay-Per-View was a vertical market, and "high-speed internet" meant expensive 128Kbps ISDN or something only businesses used – a T1. We were just beginning to scratch the surface of what was possible with digital distribution, and the infrastructure to support it didn't exist. We had to build it from nothing.

And yet, here we are. Streaming isn't just dominant; it has redefined the entertainment industry. The very idea of owning media is becoming obsolete. Not because people stopped valuing it but because access has replaced ownership as the new standard.

But this evolution has come with trade-offs. The same technology that gave consumers the freedom to listen, watch, and read whatever they wanted, whenever they wanted, also took away control over permanence. Media that was once static, immutable, and owned now exists at the discretion of corporations and licensing agreements. The version of a movie, a song, or a book that exists today may not be the same version that exists tomorrow. Who makes that decision may depend on the whims of the then-dominant culture or corporate sensibilities. Some content simply disappears. Others are quietly altered.

Which brings us back to the question: if we no longer own our media, who does? And more importantly, who decides what we get to see, hear, and remember?

The on-demand generation was born out of a desire for greater accessibility and freedom, but that same freedom now comes with new limitations. The trade-off isn't necessarily bad. It's just different. What started as an effort to eliminate the constraints of linear programming has now placed us in a world where digital curation and corporate interests shape our cultural memory.

In the end, the real lesson isn't about technology or platforms. It's what I wrote in *Website Sound* all those years ago:

"The winners here will be those entities that deliver the most compelling content, to the broadest net audience, over the most comprehensive and reliable Internet distribution system."
– Patrick Seaman, Website Sound (Circa 1996, New Riders/ McMillan)

Content was king then. It's king now. The only difference is that consumers don't hold the keys anymore.

As we move further into an era where access replaces ownership, we need to ask ourselves: Who controls the content we consume? The transition to digital has given us more choices than ever, but it has also made us dependent on platforms that can alter, remove, or restrict access at any time. If we don't preserve the original versions of our media, who will?

What are your experiences and concerns about the future of digital media? Are we trading convenience for control? Is there a way to balance accessibility with preservation?

- Do you still buy physical media, or have you gone fully digital?
- Have you ever noticed a favorite movie, song, or book altered or removed from a platform?
- How do we ensure cultural preservation in the streaming age?

4: Broadcast.com's International Legacy

Streaming and the Globalization of Media

Long before streaming became an everyday part of life, we were out there proving that it could work, not just in the U.S. but around the world. The internet wasn't just a new distribution method; it was a great equalizer, breaking down barriers that had long kept content locked within borders. If you had the right strategy (and a high tolerance for legal battles), you could bring broadcasters into the digital future.

I saw this firsthand when I worked with the BBC World Service to bring them online. At the time, much of their programming was delivered via shortwave radio and other legacy methods, but they knew the internet was the future. The challenge? Making it happen. It wasn't just about solving the technical hurdles, at least those we could engineer around. The real fight was legal. The BBC had about 300 lawyers on their side, and it was just me and Belinda Johnson, our brilliant in-house general counsel (who unsurprisingly later went on to become Airbnb's COO). I went back and forth to London for a year, hammering out agreements and making sure Broadcast.com could deliver the BBC World Service to a global audience. It wasn't easy, but we got it done, and I have to say, the people at the World Service were a class act, and lovely to work with.

At the same time, I was also advising Voice of America on how to extend its reach beyond traditional broadcasting and ham radio.

Whether it was VOA or BBC, the mission was the same: use the internet to break free from outdated distribution models and connect with audiences everywhere.

Of course, news organizations weren't the only entities eager to tap into the emerging pos   sibilities of streaming technology. Our partnerships spanned a diverse and vibrant array of industries, each pushing the boundaries of what streaming could achieve. Take our collaboration with the Fashion Channel in Paris – this dazzling partnership clearly illustrated how streaming was rapidly extending its influence far beyond the traditional realms of sports coverage and talk radio, ushering in a new era of entertainment, culture, and global connectivity.

Beyond fashion, there were countless unforgettable live events – energetic concerts featuring legendary bands like Heart, groundbreaking digital experiences such as the Playboy Cyberparty, and spectacular showcases like the *Victoria's Secret* Fashion Show. Each event was unique, expansive, and captured enormous global attention. Unsurprisingly, events like *Victoria's Secret* were bandwidth-busting mega events for us.

I will note that, as a married man at the time, I sent our young guys out to those. I did receive a token consolation gift of a signed Playboy from the women who were there while I remotely produced the broadcast.

Intel, Premiere Radio Networks & Art Bell

At one point, the radio station network carrying the late Art Bell's famous late-night program, *Coast to Coast AM*, approached us about streaming his show. Back then, Art Bell wasn't just a late-night host – he was one of the top-rated radio personalities nationwide, with the show syndicated on over 500 stations and drawing an estimated 10 to 15 million listeners each night during its peak in the late 1990s.

Making this happen required setting up equipment at Art's personal studio in Pahrump, Nevada, a small desert town arguably best known at the time for its proximity to several legal brothels.

Given Art's prominence and the critical importance that our relationship with Premiere Radio Networks was, I decided it was best if I personally oversaw the equipment setup. After calling Art directly for directions. Remember, this was the late 1990's – we didn't have smartphones and GPS navigation back then. It was paper maps.

I was amused, a bit worried, and a little taken aback, to receive instructions involving landmarks like, "Turn right at the first brothel, then left at the second one..."

Upon arriving in Pahrump after the lengthy desert drive from Las Vegas, I realized I urgently needed to find a restroom before arriving at Art's home. With the directions I'd been given, I didn't see any gas stations or other options. So, I found myself nervously stopping at the last brothel on the directions and awkwardly asking if I could use the bathroom, offering to buy a drink or pay if needed, for the inconvenience. Thankfully, given how remote everything was, this was apparently not unusual, and I quickly went on my way, feeling both relieved and amused by the surreal moment.

Art's studio was located in what he affectionately called his "bunker." As he warmly welcomed me and proudly showed off his setup, he made a point of highlighting the numerous remote security cameras monitoring the perimeter. I couldn't tell if this was a friendly demonstration or a subtle reminder of his vigilance. Art was, by all accounts, an incredibly unusual yet undeniably charming, if rather strange and peculiar personality.

My trip to Art Bell's home was the culmination of extensive negotiations involving Intel's John Kirby, Art's radio network, and Broadcast.com. Our goal was ambitious – to put cameras inside his studio, allowing us to webcast his show to a global audience. When we finally began streaming, the response was incredible. Even though broadband internet was still rare and smartphones were nonexistent, we attracted over half a million listeners to our RealAudio streams in just the first week, along with more than a hundred thousand views of the streaming video feed. By today's standards, these numbers might seem modest, but for 1999, when most listeners were still on dial-up connections, this was a significant milestone.

When Art returned our equipment at the conclusion of our agreement, his approach left our team genuinely alarmed. The gear arrived sealed inside a foil-lined box, suspiciously encased within layers of license plates. an unconventional choice that immediately set off alarm bells among our technical staff. They stood hesitantly around the mysterious package, debating whether to call in the bomb squad before daring to open it. Perhaps Art was simply being overly cautious. but his eccentric packaging certainly put everyone on edge. Despite his peculiar methods,

Looking back, I'm reminded of the pioneering spirit and eclectic characters who shaped the early days of Internet broadcasting. Although I only met Art Bell in person that one time, the experience left a lasting impression. Famous for late-night tales of 'black helicopters,' UFOs, and conspiracy theories, like his gripping 1997 interview with alleged Area 51 insider Bob Lazar, his deep, soothing voice and thoughtful style captivated millions. Broadcasting from his bedroom studio in the Nevada desert, Bell didn't shout or sensationalize; he listened, probed, and made the bizarre feel real, drawing an estimated 10 to 15 million listeners nightly at his peak in the late 1990s. His storytelling turned truck drivers, night owls, and insomniacs around the world into devoted fans of the unexplained.

Broadcast.com Japan

Another ambitious initiative was Broadcast.com Japan – a monumental project that placed me directly into the heart of strategic discussions in Tokyo, sitting in the boardroom alongside industry titans like SoftBank's Masayoshi Son. I vividly recall his enthusiasm as he proudly demonstrated his Sony Aibo robot dog, a symbol of innovation at a moment when digital media was reshaping global communication. Navigating Japan's intricate and unique media landscape was a profound experience and underscored just how transformative our work was becoming worldwide.

Then there were the live events. Our reach extended far and wide, covering events from South Africa to Asia, from major European capitals to remote locales that had never before experienced live internet broadcasts. Each event was a first, each a reminder of how, even long after Broadcast.com, I kept working at the intersection of technology and media. In later years, I launched linear TV and on-demand OTT projects in Mexico and the Philippines, proving again that the demand for streaming content isn't just global. It is inevitable.

This chapter isn't just about the expansion of streaming. It's about how it fundamentally changed the way media crosses borders. From global sports broadcasts to Netflix pushing foreign-language series into the mainstream, streaming has transformed how people consume content, breaking down barriers that once seemed impenetrable. But that transition wasn't automatic. It took pioneers willing to fight for access to convince broadcasters, content owners, and regulators that the old ways weren't just outdated. They were obsolete.

That's the story of streaming's globalization.
It's a story of opportunity, of resistance, of
massive cultural shifts. And, even today, all these
years later, it's just getting started.

Global Reach of Sports and Entertainment

Before streaming, access to international content was limited to regional broadcasts or imported DVDs, often delayed and localized for specific markets. Streaming platforms revolutionized this by offering real-time or near-real-time access to sports and entertainment worldwide.

For example:

- **Sports:** Services like DAZN, ESPN+, and Amazon Prime now bring international leagues, like the English Premier League and the NBA, to fans across continents. The ability to watch live games and highlights have grown global fanbases for teams and players who were once regional heroes. According to Statista, global sports streaming revenues reached $22 billion in 2022, a testament to this broadening reach https://www.statista.com/
- **Entertainment:** Netflix's global strategy includes licensing and producing content like *Money Heist* (Spain) and *Squid Game* (South Korea), turning localized productions into global phenomena. A report from Parrot Analytics showed that non-English series accounted for 40% of Netflix's most-watched shows in 2021 https://www.parrotanalytics.com

Cultural Exchange Through Streaming

Streaming platforms have become a modern-day bridge between cultures, allowing audiences to explore stories and traditions from around the world. Subtitles and dubbing have made foreign-language content more accessible, encouraging cultural curiosity. For example:

- The rise of K-dramas and Bollywood films on platforms like Netflix and Amazon Prime has introduced millions to South Korean and Indian cultures. *Gyeonseong Creature*, from South Korea on Netflix, is one example of a show I would have never discovered otherwise. Hollywood – you could learn a thing or twelve.
- Documentaries like *The Act of Killing* and *My Octopus Teacher* have fostered international dialogue on social justice and environmentalism.

Building International Fandoms

Streaming has enabled fans from different countries to connect and celebrate shared interests. Social media amplifies this by creating virtual communities where fans of shows, sports teams, or influencers can engage in real time.

For example:

- The international fanbases of sports stars like Cristiano Ronaldo or Serena Williams thrive on streaming and social media platforms, which broadcast their games and amplify their personal brands.
- Fan campaigns for shows like *Dark* (Germany) and *Stranger Things* (U.S.) have sparked collaborative global discussions, memes, and even fan-made content.

Challenges and Opportunities

While streaming has expanded cultural access, it also raises questions about content localization and cultural homogenization. Platforms must strike a balance between preserving cultural integrity and making content relatable for global audiences. Services that embrace this complexity, like offering multi-language options or region-specific recommendations, are poised to lead in the next phase of globalization.

Why Some Streaming Startups Thrive While Others Struggle: The Global Battle for Viewers

In May 1998, Intel invited me to speak at Beijing University about, well, the birth of the Internet and streaming media. I still remember standing in that lecture hall, looking out at an audience that was absolutely locked in. Wide-eyed, leaning forward, hanging on every word. The thirst for knowledge, the sheer hunger to understand this new frontier, was palpable. Despite the language barrier and the translator's admirable efforts, the Q&A session stretched long past its scheduled time. They wanted to know everything. How streaming worked, where it was headed, and how they could be part of it. I was deeply moved by the experience.

And it wasn't just in Beijing. From New York to Dallas, San Jose to Tokyo, Paris to London, the energy in those early days was intoxicating. Everywhere I went, there was the same exuberant optimism, a sense that we were building something world-changing, something that would connect people in ways never before possible. The enthusiasm was never in question.

So why, after all that global momentum, have some regions exploded in streaming growth while others have lagged behind? The answer lies not in passion because that was universal but in the unique mix of market forces, regulatory environments, infrastructure, and cultural adaptation that have shaped the industry in each part of the world. Some markets cleared the runway for streaming to take off, while others erected roadblocks that made scale and sustainability far more difficult.

And that's what we'll dive into next.

Market Dynamics and Competition

In regions where the market is saturated with dominant players like Netflix, Amazon Prime Video, and Disney+, smaller streaming services often find it challenging to carve out a niche. The substantial capital and extensive content libraries of these giants create high entry barriers for newcomers. Conversely, in emerging markets or regions with less penetration by major platforms, local streaming services have the opportunity to thrive by offering tailored content that resonates with regional audiences. For instance, platforms like iQIYI in China and Viu in Southeast Asia have gained significant traction by focusing on local content and understanding regional viewing habits.

Regulatory Environments

Government policies and regulations play a pivotal role in shaping the success of streaming businesses. In some countries, stringent content regulations, licensing requirements, and censorship laws can pose challenges for both international and local streaming services. However, these same regulations can also provide a protective buffer for domestic platforms against foreign competition. For example, France's cultural policies require streaming platforms to prioritize European content, invest heavily in local productions, and maintain quotas that limit foreign dominance. These measures have strengthened domestic services like Canal+ and Salto while ensuring French media retains its cultural identity in the streaming age.

Technological Infrastructure

The quality and accessibility of technological infrastructure, particularly internet connectivity and broadband penetration, are critical determinants of a streaming service's viability. In regions with advanced infrastructure, consumers expect high-definition streaming with minimal buffering, necessitating significant investment from service providers. In contrast, areas with limited bandwidth or geographical challenges may require platforms to optimize content delivery and offer lower-resolution options to ensure a seamless user experience.

For example, when I worked on a streaming project in the Philippines at the former Clark Air Base, I encountered firsthand how geography and history shaped technological challenges. After Mount Pinatubo's catastrophic eruption in 1991, the base was buried under volcanic ash, leading to its abandonment by the US military. The Philippine government later transformed this site into an international airport and free trade zone. Our operations were housed in the facility that had once served as the base's data center, now repurposed as a TV/cable network headquarters filled with antiquated equipment.

The Philippines' archipelagic geography made traditional cable TV expansion nearly impossible across its 7,641 islands. Cable providers

had turned to satellite broadcasting as an alternative, but faced rampant piracy through knockoff satellite dishes that severely undercut legitimate subscription revenues. Our solution was elegantly efficient: we replaced rows upon rows of aging equipment with just three racks of modern hardware that ingested satellite feeds, encoded 50 linear channels, and securely broadcasted them via internet-based OTT distribution. This approach effectively circumvented both the physical limitations of cable infrastructure and the piracy vulnerabilities of conventional satellite broadcasting.

This kind of adaptability to technological constraints is crucial for streaming platforms looking to succeed in diverse markets, whether it's optimizing content for low-bandwidth regions, securing content delivery against piracy, or developing alternative distribution methods where traditional infrastructure is impractical. Our work at Clark demonstrated how streaming technology could leapfrog legacy  systems in developing markets, providing secure, scalable solutions that traditional broadcasting simply couldn't match.

Cultural Preferences and Content Localization

Cultural nuances and content preferences vary widely across regions. Streaming services that invest in content localization, such as subtitles, dubbing, and culturally relevant programming, are more likely to engage local audiences. Moreover, platforms that produce or acquire content reflecting local stories, languages, and traditions can build a loyal subscriber base. For instance, India's Hotstar (now Disney+ Hotstar) achieved some success by offering a mix of local television shows, movies, and sports content, catering specifically to Indian viewers' tastes.

Economic Factors

Subscription pricing strategies must align with the economic realities of the target market. In regions with lower disposable incomes, streaming services that offer flexible pricing models, such as ad-supported free tiers or affordable subscription plans, are more likely to attract and retain subscribers. Additionally, partnerships with local telecom providers for bundled offerings can enhance affordability and accessibility, further driving adoption.

In conclusion, the success of small streaming businesses is intricately linked to their ability to navigate and adapt to regional market conditions, regulatory landscapes, technological infrastructures, cultural preferences, and economic factors. By tailoring their strategies to these local nuances, smaller streaming platforms can find fertile ground to grow and compete, even in the shadow of global giants.

The Content Wars: Locked Libraries and the Battle for Rights

In the streaming business, technology is just the delivery mechanism. **Content is the currency.** From the very beginning, we knew that no matter how innovative our platform was, it meant nothing if we didn't have something compelling to stream. That hasn't changed. If anything, the battle for content has only intensified.

For today's streaming startups, the biggest challenge isn't building the tech. Cloud services, CDNs, and off-the-shelf streaming solutions have made that part easier than ever.

The real fight is for content.

And in many cases, that fight is already over before a new player even steps into the ring.

The Lockdown: Who Owns What?

In the early days, content rights were fragmented, and getting them was half the battle. At AudioNet, I'd guess that about half of our first twenty hires were attorneys because even from the beginning, we knew that securing the rights to stream content was just as critical as the technology itself. We made deals directly with radio stations, sports teams, and event organizers. We knocked on doors, pitched the vision, and got agreements in place. In every case, negotiating with organizations whose existing contracts had no provision for the Internet as a distribution method. We were writing the rulebook as we went.

I often spent my days on the phone explaining what we were doing to rights holders and my evenings on the tech, working with our small but effective engineering team.

The Days of Signing up "New" Content/
Rights-Holders are long gone.

Today, the vast majority of premium content, including movies, TV shows, and major sports rights, are already locked up in exclusive, long-term licensing agreements. A handful of major players, including Disney, Warner Bros. Discovery, Netflix, Amazon, and Apple, own or control an overwhelming share of the world's most valuable content libraries. If you want access to those libraries, you're either paying a fortune, or you're out of luck.

This is why we've seen aggressive content consolidation. Disney bought 21st Century Fox. Warner merged with Discovery. Amazon swallowed MGM. Apple is pouring billions into exclusive content. Netflix isn't just licensing anymore. It's producing at a scale that rivals Hollywood. Everyone who could afford to lock up content has done exactly that.

The Content Starvation Problem

For new entrants in the streaming space, this presents a huge problem because if you don't own content, you have to license it. And if the best content is already tied up in long-term deals, what's left?

That's why smaller platforms often struggle to differentiate. Many end up fighting over the same scraps. Older catalog content, niche genres, or regional programming. Some find success by going deep instead of wide, focusing on anime (Crunchyroll), classic films (The Criterion Channel), or international dramas (Viki). Others rely on user-generated content (YouTube, TikTok, Twitch), where the audience becomes the creator.

But for most, the reality is simple:

If you don't control your content, you don't control your future.

The Price of Exclusivity

Even among the big players, the arms race for exclusivity has created a financial sinkhole. Platforms have poured billions into acquiring content libraries, locking in exclusive deals, and outbidding each other for new originals. The result? A market where content costs have skyrocketed, subscriber growth is slowing, and profitability remains elusive.

Netflix spends more than $17 billion per year on content. Amazon: $23 billion, Apple: $20 billion, and Disney reportedly $23.4 billion. Meanwhile, smaller platforms are getting squeezed out, unable to compete with these staggering budgets. But spending billions doesn't guarantee success. Amazon sank well over a billion dollars into Rings of Power, only to see it fail spectacularly in audience engagement. Disney has poured billions into new shows, many of which, to be very generous, have struggled to become blockbusters. These financial missteps highlight a brutal reality: money alone can't buy a loyal audience.

And as major studios pull their content back to fuel their own streaming services, like when NBCUniversal yanked The Office from Netflix, or HBO took Friends away from competing platforms, the fight for dominance gets even bloodier. The result? More fragmentation, more competition, and a growing number of streaming services scrambling for a shrinking pool of paying subscribers.

The Way Forward for New Players

So, where does that leave new or smaller streaming businesses?

- Original Content or Bust – If you can't license top-tier content, you have to create your own. However, producing original content is expensive and high-risk, and it takes time to build an audience. Even a breakout hit doesn't guarantee sustainability. Just ask Quibi.
- Niche and Specialized Markets – Some platforms find success by focusing on underserved genres or communities. Horror (Shudder), British TV (BritBox), and sports documentaries (DAZN) are all examples of carving out a space by going after specific audiences.
- Live and Interactive Content – Streaming services that embrace live events, interactive experiences, or community-driven content can break through. Twitch and YouTube thrived on this model. Newer players are experimenting with live shopping, audience participation, and gamified content experiences.
- Hybrid Business Models – The days of a single subscription model ruling all are fading. Ad-supported tiers, FAST (Free Ad-Supported Television) services, and hybrid monetization models are giving platforms more flexibility in how they grow. Even Netflix and Disney+ are now embracing ads.

The Content Lock Will Define the Next Decade

Ultimately, the battle for streaming isn't just about tech. It's about who controls the stories we watch. The platforms that own their content or can afford to license at scale will continue to dominate. The ones left scrambling for leftovers will have to get creative or get crushed.

The streaming wars aren't over. They're just entering their next phase. And for smaller players, the only real path to survival is finding content that people will pay for and that the big guys don't already own.

Closing Summary for Chapter 4: The Globalization of Streaming Media

Streaming has done more than just change how we consume content. It has reshaped cultural exchange, erased geographical barriers, and redefined global media distribution. What once required satellite deals, licensing negotiations, and physical distribution now happens instantly, in real time, with the tap of a screen.

From BBC World Service embracing the internet to Netflix turning foreign-language series into global blockbusters, streaming has blurred the lines between local and international content. Sports leagues have found massive new audiences, K-dramas have become household names in the West, and indie filmmakers from any country can now reach millions without ever securing a traditional distribution deal.

But this shift hasn't come without challenges. As streaming centralizes control of content, who decides what gets shown and what doesn't? What happens when licensing deals restrict access to media across regions? And how do we ensure that globalized content doesn't lead to cultural homogenization, where smaller voices are drowned out by the biggest players?

The next phase of global streaming won't just be about expanding access. It will be about who controls the platforms, who owns the content, and who decides what the world gets to watch.

What's Next?

The globalization of streaming is still evolving, and the conversation isn't over. What do you think?

- Do you prefer globalized content, or do you seek out regional and niche streaming services?
- Have you ever struggled to access content due to licensing restrictions in your region?
- What do you think is the next frontier for streaming? More global content or more fragmentation?

5: The Internet Broadcasting Boom

A Competitive Landscape

In the mid to late 1990s, the internet was rapidly evolving, giving rise to various platforms that sought to capitalize on the burgeoning digital media landscape. Broadcast.com emerged as a pioneer in internet radio and streaming, but it was not without competition. Several companies and technologies vied for dominance in the online broadcasting arena, each adopting unique approaches to content delivery and monetization.

Broadcast.com and the Birth of Internet Radio

I was hired by Mark Cuban and Todd Wagner before the name change to AudioNet in 1995. Later, after we were able to start streaming video as well as audio, we changed the name to Broadcast.com. The platform aggregated a variety of media, including live sports broadcasts, radio stations, and corporate events, making it the most comprehensive source for streaming content on the internet. Its successful initial public offering in 1998 and subsequent acquisition by Yahoo! in 1999 for $5.7 billion underscored its prominence in the market. https://en.wikipedia.org/wiki/Broadcast.com

Napster: The Disruptor That Changed Music Forever

Launched in 1999 by Shawn Fanning and Sean Parker, Napster introduced a peer-to-peer (P2P) file-sharing model that allowed users to share and download music files directly from each other's computers. Unlike Broadcast.com's streaming model, Napster facilitated the distribution of MP3 files, enabling users to build personal music libraries without purchasing physical media. This approach disrupted the traditional music industry, leading to significant legal challenges and its eventual shutdown in 2001. https://www.avclub.com/for-better-or-worse-napster-changed-the-music-industry-1844708800

RealNetworks: The Technology Behind Early Streaming

Founded in 1994, RealNetworks developed RealAudio and RealVideo, among the first technologies to enable streaming media over the internet. The company provided tools for both live and on-demand streaming, catering to broadcasters and content creators aiming to reach online audiences. While RealNetworks focused on developing the underlying technology for streaming, Broadcast.com concentrated on aggregating and delivering diverse content to users. Real was the dominant streaming platform until Microsoft released its competing platform(s).

RealNetworks began losing its dominant position in the streaming industry in the early-to-mid 2000s, as several key factors contributed to its decline:

Key Events Leading to Real's Decline:

1. 2005-2007: Rise of YouTube and Flash Video:
    o YouTube launched in 2005, providing a free and easily accessible way to stream video content. It quickly

surpassed RealPlayer as the preferred platform for online video streaming.

- o Adobe Flash Video (FLV) became the standard for web-based streaming, making Real's proprietary formats (RealAudio and RealVideo) less relevant.

2. 2007-2008: The Rise of Netflix and Silverlight

- o Netflix launched its streaming service in 2007, using Microsoft's Silverlight technology instead of Real's format.
- o Apple and Microsoft increasingly favored their own media players (QuickTime and Windows Media Player) over RealPlayer.

3. Late 2000s: Decline in Consumer Adoption

- o RealNetworks' freemium model became less appealing as competitors offered free, ad-supported streaming.
- o RealPlayer became known for bundled software and intrusive updates, pushing users toward simpler, more seamless options like VLC, Windows Media Player, and QuickTime.

4. 2010s: Final Shift Away from RealNetworks

- o HTML5 video became the new web standard, eliminating the need for proprietary streaming formats like RealVideo.
- o By the early 2010s, RealNetworks had completely lost its dominance, shifting focus toward enterprise solutions and niche streaming services rather than consumer media players.

Final Turning Point:

By 2010, RealPlayer was largely irrelevant in the mainstream consumer streaming market, with YouTube, Netflix, and cloud-based services taking over.

NetRadio: The Rise and Fall of Internet-Only Broadcasting

Established in 1995, NetRadio positioned itself as one of the first internet-only radio networks. It offered a variety of genre-specific channels, providing continuous music streams without traditional broadcast constraints. Unlike Broadcast.com, which included a mix of live events and rebroadcasts, NetRadio focused exclusively on music programming, targeting niche audiences seeking specific genres.

We later bought them.

Clash of Models: Centralized Platforms vs. Peer-to-Peer Networks

- **Content Delivery**: Broadcast.com aggregated existing media content, including live events and radio station feeds, providing a centralized platform for users. In contrast, Napster enabled decentralized file sharing among users, facilitating the exchange of music files without a central repository.
- **Monetization**: Broadcast.com generated revenue through live event streaming services, advertising, and partnerships, leveraging its platform to deliver ads alongside streaming content. Napster, however, operated without a clear revenue model, offering free access to shared music files, which contributed to its legal challenges.
- **Legal and Ethical Considerations**: Broadcast.com secured rights to stream content, working within existing legal frameworks. Napster's model, however, raised significant copyright infringement issues, as it facilitated the unauthorized distribution of music, leading to lawsuits from the recording industry.

While Broadcast.com and its contemporaries like RealNetworks and NetRadio operated within the bounds of emerging digital media regulations, Napster's disruptive approach highlighted the tension between technological innovation and intellectual property rights. These early competitors each contributed to the evolving landscape of internet broadcasting, setting the stage for the diverse streaming platforms that dominate today's media environment.

Streaming Wars: Evolution from Early Internet Broadcasting to Today's Competitive Landscape

The term "streaming wars" describes the intense competition among streaming platforms vying for market share in the digital entertainment industry. While today's streaming wars involve major players like Netflix, Disney+, and Amazon Prime Video, this battle for audience attention has its roots in the early days of internet streaming. Examining the parallels and differences between these periods offers insight into the industry's evolution.

Early Internet Streaming: The Pioneering Era

In the mid-1990s, companies like Broadcast.com, RealNetworks, and iCraveTV were at the forefront of internet broadcasting. These pioneers faced challenges such as limited bandwidth, rudimentary streaming technology, and a nascent user base unfamiliar with consuming media online. The competition was primarily about technological innovation and establishing proof of concept.

For instance, Broadcast.com aggregated various media content, including live sports and radio broadcasts, to provide a comprehensive streaming experience. RealNetworks developed early streaming codecs, enabling audio and video playback over the internet. Real tried to compete directly with Broadcast.com but, in my opinion, was a divided company. They couldn't decide whether to be a platform or a broadcaster. It was an internal schism that hurt them.

iCraveTV attempted to stream over-the-air television broadcasts online, raising significant legal questions about content rights. Dan Rayburn:https://www.streamingmediablog.com/2016/03/history-of-the-streaming-media-industry.html

During this era, the focus was on overcoming technical limitations and convincing both content creators and consumers of the viability of online media distribution. The market was fragmented, with each player experimenting with different models to attract users and secure content partnerships.

The Modern Streaming Wars: Content is King

Today's streaming landscape is characterized by a battle for exclusive content, subscriber retention, and global expansion. Major corporations have launched their own platforms, leading to a crowded market where content libraries and original programming are key differentiators. For example, Disney+ leverages its extensive catalog of beloved franchises, while Netflix invests heavily in original content to attract and retain subscribers. https://race.com/blog/history-of-streaming-services/

Unlike the early days, when technological feasibility was the primary concern, the current competition centers on content acquisition and production. The introduction of exclusive shows and movies has become a strategic move to draw subscribers. Additionally, the business models have evolved, with platforms exploring various monetization strategies, including ad-supported tiers and bundled services. https://insights.som.yale.edu/insights/how-the-streaming-wars-will-alter-the-media-landscape

Key Differences Between Then and Now

- **Scale and Investment**: Early streaming services operated on limited budgets, focusing on niche audiences and experimental content delivery. In contrast, today's platforms are backed by substantial financial resources, enabling large-scale content production and global marketing campaigns.
- **Technology and Infrastructure**: Advancements in internet

infrastructure, such as high-speed broadband and content delivery networks, have mitigated many of the technical challenges faced by early streamers. This allows for high-definition streaming and a seamless user experience, which were unattainable in the 1990s.

- **Market Saturation and Consumer Choice**: The current market is saturated with numerous streaming options, leading to intense competition for subscriber dollars. Consumers now face decisions about which services to subscribe to, often based on exclusive content offerings, whereas early users had limited options and were primarily early adopters of new technology.
- **Legal and Licensing Challenges**: Early services like iCraveTV navigated uncharted legal territories concerning content rights, sometimes leading to shutdowns due to copyright infringement issues. Today's platforms operate within well-defined legal frameworks, often securing exclusive rights and producing original content to avoid such pitfalls.

Speaking of content rights, it's worth emphasizing just how crucial – and challenging – they were during the early days of AudioNet and Broadcast.com. To give you some perspective, about half of our initial hires – roughly six of the first dozen employees – were attorneys. This wasn't an accident. At the time, existing legal frameworks and contractual agreements simply didn't anticipate the internet or streaming as viable distribution channels. Every deal we made had to break new legal ground. Our legal team didn't just finalize agreements; they were inventing entirely new kinds of contracts, paving the way for a medium that had never existed before. They weren't simply negotiating rights; they were defining them.

One particularly vivid example was our collaboration with the BBC World Service. Bringing them online meant much more than solving technical challenges. It required navigating one of the world's most complex legal and regulatory landscapes. For nearly a year, I shuttled back and forth to London, often spending entire days negotiating with the BBC's formidable legal team, approximately 300 attorneys strong.

Fortunately, I had the steadfast support of our brilliant in-house counsel, Belinda Johnson, who later became widely recognized for her role as COO of Airbnb. Together, we patiently worked through countless drafts, clauses, and revisions, carefully aligning each detail to create a precedent-setting agreement. It was a marathon, but our persistence paid off: we successfully completed the deal, enabling the BBC World Service to leverage our technology and stream globally for the first time.

BBC World Service and Princess Diana

In the wake of the tragic and untimely death of Princess Diana, we made the decision to carry the funeral live, no paywall, no subscription, just a raw public service broadcast. It was a gesture of solidarity and respect, done as a favor to our colleagues at the BBC World Service, who were scrambling under the weight of global attention and a sudden, historic moment.

Back then, the technology was still early, rudimentary by today's standards, but powerful in the right hands. We ran the stream through our servers and then pushed it out to a rapidly growing internet audience. And because of how our infrastructure was set up, one of the first to truly understand and track global streaming at scale, we could see, in real time, where the viewers were tuning in from.

What we saw was nothing short of fascinating.

This wasn't just the general public mourning a princess. We saw government offices lighting up across the map. Secure buildings. Embassies. Ministries. Places that don't typically show up on civilian audience metrics. From Washington to Moscow, from Riyadh to Pretoria, the digital trace of this singular global moment unfolded on our dashboards, like a world holding its breath in unison.

It was one of those times where the emotional gravity of an event and the technical reach of our platform intersected. We weren't just streaming a funeral. We were watching history ripple across the globe, packet by packet, from palace to parliament.

And in that moment, we understood the true power of this new medium, not just to entertain or inform, but to connect humanity in real time, across borders and ideologies, with a shared sense of loss and reverence.

That was one of the first times I realized we were sitting on more than just a delivery system. We were holding the beginnings of a global nervous system.

Summary

In summary, while the early days of internet streaming were marked by technological experimentation and the challenge of proving the concept, today's streaming wars revolve around content dominance, subscriber acquisition, and retention strategies. The evolution reflects a shift from pioneering technology to leveraging vast content libraries and original productions to capture audience attention in a highly competitive market.

Legacy Impact: Broadcast.com's Influence on Modern Streaming Platforms

Broadcast.com launched in 1995, was a trailblazer in the realm of internet broadcasting, pioneering practices and technologies that have significantly influenced today's leading streaming platforms such as YouTube, Spotify, and Netflix.

Pioneering Live Streaming

Broadcast.com was among the first to stream live events over the internet, including sports games and concerts. This practice laid the groundwork for platforms like YouTube, which now offers live-streaming features for events ranging from gaming tournaments to global concerts. The concept of delivering real-time content to a worldwide audience, initiated by Broadcast.com, has become a standard expectation in modern streaming services.

Aggregation of Diverse Content

By aggregating various media types, such as radio broadcasts, live sports, and corporate events, Broadcast.com provided users with a centralized platform for diverse content consumption. This model is evident in platforms like Spotify, which offers a vast library of music, podcasts, and live audio sessions, and Netflix, which provides a wide array of movies, TV shows, and documentaries. The strategy of content aggregation to cater to varied user interests can be traced back to Broadcast.com's approach.

User Engagement and Interactive Features

Broadcast.com incorporated interactive elements, such as chat rooms and user forums, to enhance user engagement during live broadcasts. This focus on community building and real-time interaction is mirrored in modern platforms like YouTube, where live chats accompany streams, and Spotify, which integrates social features allowing users to share playlists and see friends' listening activities. The emphasis on fostering a sense of community among users has its roots in Broadcast.com's early initiatives.

Monetization Strategies

Broadcast.com utilized paid streaming services for corporations for live events, advertising, and sponsorships as primary revenue streams, setting a precedent for monetization in digital media. Today, platforms like YouTube employ similar models, offering free access supported by advertisements while also providing premium, ad-free subscription options. This multifaceted approach to monetization reflects the foundational strategies implemented by Broadcast.com.

In summary, Broadcast.com's early adoption of live streaming, content aggregation, user engagement features, and monetization strategies has left a lasting legacy, shaping the functionalities and business models of contemporary streaming platforms.

Beyond Consumer Platforms: The Evolution of Internal Corporate Streaming

The Bell System, often referred to as "Ma Bell," was a system of companies led by AT&T that dominated telephone services in North America for over a century. When the Bell System was broken up in 1984 due to an antitrust lawsuit, it resulted in the creation of seven independent Regional Bell Operating Companies (RBOCs), also known as "Baby Bells"2.

The seven "Baby Bells" were:

- **Southwestern Bell Corporation** (later SBC Communications, Inc., now part of AT&T Inc.)
- **Bell Atlantic** (later merged with GTE to form Verizon Communications)
- **NYNEX** (merged with Bell Atlantic to form Verizon Communications)
- **Pacific Telesis** (later merged with SBC Communications, Inc.)
- **BellSouth** (later acquired by AT&T Inc.)
- **Cincinnati Bell** (remained independent)
- **US West** (later merged with Qwest Communications, which was then acquired by CenturyLink)

After the SBC, Pacific & BellSouth merged back into and became the current AT&T, they used their internal infrastructure more and more to communicate internally. Each of the old entities had their own studio. They naturally adopted streaming as a powerful tool to coordinate with their (80k-ish, if I remember) employees. Not only did they use streaming, but they used it aggressively. So much so that their production facilities couldn't keep up with demand.

I was brought in as an outside subject matter expert to advise them on how to improve efficiencies. As I dug into the problems, I found that their studios were hopelessly outdated. Their engineers were literally **soldering resistors on circuit boards** just to keep their aging cameras operational. Despite their extensive internal network for employee webcasting, they were relying on decades-old technology that was rapidly becoming unsustainable.

A major pain point was production efficiency, or rather, the lack of it. Their existing studios were too slow to set up for each shoot, making it difficult to produce internal content at the speed they needed. What should have been a rapid, agile process was instead bogged down by inefficient workflows and outdated gear. They required orders of magnitude improvements in production efficiency, and their existing infrastructure simply couldn't keep up.

One of the projects involved included building a new central studio at their headquarters, a highrise in Dallas. The studio would be on one of the upper floors. If any of you have experience with studios, you'll immediately see an… issue. Broadcast studios traditionally have special needs for space and infrastructure that are not… well suited… for an (old) highrise building.

I'll give you one example issue: Lighting. They had planned to use traditional studio lighting. Now, any of you who have worked in a traditional studio know about the HVAC issues. Those lights are ***hot*** and require lots of power and HVAC for cooling. Simply running all that HVAC can require you to make structural changes. This adds up to a lot of expense.

What shocked me was how, even though LED studio lighting was "relatively" new at the time, all the major broadcast industry studios were already transitioning. That said, I still had to convince the powers that be to consider it. I brought in an outside studio lighting expert to do a "shoot out" doing a demonstration production to compete with their older legacy equipment.

However, lighting was only one piece of the puzzle. They needed a full reset of their production ecosystem, integrating modern cameras, production equipment, and computer systems to streamline the workflow in order to achieve the studio turnaround time needed to meet their rapidly growing production needs.

Despite the clear benefits, change isn't always easy, especially in large organizations. Some of their engineers were skeptical and resistant to switching from traditional studio lighting and other technology.

Unfortunately, by the time this demonstration convinced them, they had already committed to an expensive renovation, rearranging structural supports in their high-rise headquarters to accommodate old-school studio lighting instead of adapting to modern solutions.

Despite my efforts to steer them in the right direction, I was periodically called back over the years to help resolve issues that stemmed from not fully implementing my recommendations. Each time, it was another reminder that modernizing corporate broadcast infrastructure is not just about technology. It's about overcoming legacy thinking and institutional inertia.

Closing Summary for Chapter 5: The Internet Broadcasting Boom

The late 1990s were a chaotic and groundbreaking period for online media as companies raced to define what digital broadcasting could be. The early competitors, Broadcast.com, Napster, RealNetworks, and NetRadio, each approached the challenge from different angles, paving the way for today's streaming giants.

Broadcast.com demonstrated the power of aggregation and live streaming, pioneering business models that would shape the future of internet media. Napster, while legally controversial, proved that audiences craved instant access to digital content. RealNetworks established some of the earliest foundations for streaming technology, and NetRadio capitalized on the idea of digital-first broadcasting.

While the players may have changed, the fundamental battle remains the same: content, accessibility, and technology shape the success of digital media. The early internet broadcasting wars weren't just about market dominance. They were a proving ground for what would become the entertainment landscape of today.

How do you think early internet broadcasting shaped today's streaming wars? Which of these early platforms do you think had the biggest impact on modern digital entertainment?

- What do you remember about early internet streaming?
- Did you ever use Napster or early streaming platforms?
- How do you think modern services compare to those early pioneers?

6: The Evolution of Streaming

From Broadcast.com to Today's Global Audiences

How User Behavior Shaped the Streaming Industry

Streaming didn't just change how content was delivered. It fundamentally reshaped how audiences interacted with media. The shift from traditional formats to digital streaming wasn't immediate, and in the early days, people weren't entirely sure it would work.

At first, consumers were excited but skeptical. Could the internet really replace television, radio, and physical media? Bandwidth limitations, buffering issues, and poor video quality made streaming feel more like a curiosity than a true alternative to cable or DVDs. I saw this firsthand during the Broadcast.com years. People loved the idea of live, on-demand content, but they still expected the instant reliability of traditional broadcasting.

The fundamental shift from passive media consumption (traditional TV and radio) to interactive digital engagement (YouTube, Twitch, and TikTok) reshaped the streaming industry. Early pioneers like Broadcast.com proved that audiences wanted on-demand access, but it wasn't until user-generated content platforms emerged that streaming became truly democratized.

Yet, democratization did not guarantee visibility. Platforms like YouTube promised creators unlimited reach, but the reality is starkly different. As of mid-2024, YouTube hosts an estimated 14.8 billion videos, yet the median video receives just 41 views. Even more striking, 4% of all videos have zero views, meaning millions of uploads never find an audience. The notion that anyone can "go viral" is largely an illusion. One dictated by algorithms rather than pure demand. https://www.techspot.com/news/106791-youtube-numbers-uncovering-youtube-ghost-town-billions-unwatched.html

Global Streaming: By the Numbers

Back when we were expanding Broadcast.com internationally, we had virtually no data on cross-border content consumption. We were flying blind, making educated guesses about what might work in different markets. To be honest, we were mostly concerned with distributing American content to foreign markets and working to build the infrastructure to support that.

The concept of a Korean drama becoming a global phenomenon or a Spanish-language series topping U.S. charts would have seemed far-fetched to say the least. But that's exactly what's happened in recent years.

The data paints a fascinating picture of just how dramatically streaming has reshaped global media consumption. Netflix reported that in 2023, international viewing of non-English content increased by 70% compared to 2019. The platform's most-watched series of 2023, "Squid Game," drew 95% of its audience from outside South Korea, with over 142 million households tuning in during its first four weeks. That's not just impressive – it's revolutionary. Nothing in traditional broadcasting history comes close to that scale of cross-cultural viewership.

But it's not just about Korean content breaking through to Western audiences. The flows of content have become remarkably multidirectional. Spanish-language series like "Money Heist" garnered over 65 million viewers outside Spanish-speaking regions. Meanwhile, shows from France, Germany, and Japan regularly appear in Netflix's global top 10 lists, sometimes outperforming Hollywood productions in multiple markets.

Looking at YouTube, the pattern becomes even more pronounced. According to their internal metrics, the average American viewer now spends almost 40% of their watch time consuming content created outside the U.S., a figure that's doubled since 2017. Globally, over 2 billion monthly active users regularly watch content in languages other than their primary one, with subtitles or auto-translation enabling cultural discovery in ways that were unimaginable even a decade ago.

Beyond sheer viewership numbers, the economic impact is substantial. A 2024 study by Media Partners Asia revealed that international licensing and distribution of streaming content generated approximately $18.7 billion in revenue annually, creating a truly global marketplace for creators. Countries like South Korea have seen their content export values skyrocket from under $400 million in 2018 to over $1.2 billion in 2023.

Of course, these cross-border flows aren't evenly distributed. U.S. content still dominates global viewing hours (about 47% of all streaming content consumed worldwide), but that share has been steadily decreasing as platforms invest in local production hubs and viewers embrace international storytelling. The fastest-growing sources of globally successful content include South Korea, Spain, Mexico, India, and Nigeria – creating new centers of cultural production and influence.

This data tells a story we could only dream of back at Broadcast. com – streaming hasn't just connected existing content to new audiences; it's fundamentally rebalanced where culturally influential content comes from. When a series produced in Seoul, Mexico City, or Mumbai can captivate viewers in Nebraska, Berlin, and Sydney simultaneously, we've moved well beyond traditional models of cultural export. We're witnessing the emergence of the first truly global media ecosystem, one where a compelling story can find its audience regardless of where it originates.

(https://top10.netflix.com/, https://www.parrotanalytics.com/, https://media-partners-asia.com/)

The Crowdfunding Revolution: Democratizing Media Production Globally

While streaming platforms broke down distribution barriers, another revolution was quietly reshaping who could create content in the first place. Crowdfunding platforms like Kickstarter, Indiegogo, and Patreon emerged as powerful alternatives to traditional gatekeepers, letting creators go directly to fans for financing.

This shift didn't just benefit American creators – it globalized production in ways traditional funding models never could. I've watched with fascination as international artists who would have never gotten past a Hollywood executive's assistant suddenly found their visions funded by supporters across borders. Creators from around the world were suddenly able to pitch directly to audiences rather than executives.

The numbers tell an impressive story. Kickstarter alone has channeled over $7 billion to creative projects since its founding, with film and video projects receiving over $500 million of that funding. What's remarkable isn't just the dollar amount but the geographic diversity – creators from more than 170 countries have successfully funded projects, and backers contribute from virtually every nation where online payments are possible.

This global democratization of funding has had profound impacts. Take the case of "Kung Fury," a Swedish martial arts comedy that raised $630,000 from backers across the globe. The 30-minute short film became a viral sensation in 2015, eventually leading to a feature-length sequel backed by traditional Hollywood financing. The path from crowdfunded project to mainstream success has become increasingly common, creating a new talent pipeline that bypasses traditional geographic constraints.

Audiobooks and podcasts have seen similar transformations. At Timberwolf Press, we struggled to fund audio productions through traditional channels. Today, creators use platforms like Kickstarter to pre-sell audiobooks directly to listeners, often raising substantial sums for high-quality productions that would have been impossible to finance otherwise. The Audiobook Creation Exchange (ACX) further simplified global production by connecting authors with voice talent worldwide, creating a marketplace that operates across borders.

The World Bank has recognized crowdfunding's potential to democratize access to capital for creative ventures in emerging markets. Their research highlights how these platforms can provide critical funding for creative entrepreneurs in regions typically underserved by traditional financing channels. This aligns perfectly with the streaming revolution – as content from diverse cultures finds global audiences through platforms like Netflix and YouTube, crowdfunding ensures that creators from these same regions can fund new projects based on their growing international appeal.

What's fascinating about this development is how it complements the streaming revolution. Creators can now fund projects through global microtransactions, produce them with internationally distributed talent, and distribute them on borderless platforms – a complete ecosystem that exists parallel to traditional media structures. The old geographic constraints haven't just been loosened; they've been fundamentally reimagined.

Of course, crowdfunding isn't without challenges. Success often depends on creators already having some audience or marketing savvy. Many worthy projects still go unfunded, particularly from communities with limited social media reach. And fulfilling backer rewards across international borders introduces compliance headaches that would make our early days of international licensing at Broadcast.com seem straightforward by comparison.

But these growing pains don't diminish the transformative impact. When a filmmaker anywhere in the world can fund a project with support from viewers across continents, then distribute that finished film to those same audiences via streaming platforms, we've moved into territory that traditional broadcasting could never have imagined. The combination of crowdfunding and streaming hasn't just changed who can watch content – it's fundamentally altered who can create it in the first place.

- Kickstarter Stats and Facts: https://www.kickstarter.com/help/stats
- "Kung Fury" Kickstarter Success: https://www.kickstarter.com/projects/kungfury/

kung-fury
- Audiobook Creation Exchange (ACX) Global Marketplace: https://www.acx.com/help/about-acx/200484860
- World Bank Report on Crowdfunding: https://documents1.worldbank.org/curated/en/409841468327411701/pdf/840000WP0Box380crowdfunding0study00.pdf
- Encounters Documentary Festival (featuring crowdfunded films): https://encounters.co.za/

Today's digital platforms rely on algorithmic gatekeeping

Engagement statistics further reveal the challenges of breaking through in a saturated content landscape. Over 74% of YouTube videos have no comments, and 89% receive no likes. This data suggests that while the tools to create content are widely available, audience attention is far scarcer. Unlike traditional broadcasting, where programming directors curated content for mass appeal, today's digital platforms rely on algorithmic gatekeeping to determine what has surfaced and what remains buried.

Production quality also plays a crucial role in shaping user engagement. Despite the perception of highly polished content dominating YouTube, only 14% of videos feature a professional set or background, and just 38% show signs of editing. More than half suffer from shaky camerawork, and 85% have inconsistent or poor audio quality. Additionally, 40% of videos are simply music tracks with no voice-over or visual engagement, further illustrating that while streaming enables unlimited uploads, it does not ensure quality or viewership.

While early streaming platforms focused on long-form, TV-style content, the typical YouTube video in 2024 is just 64 seconds long, with over one-third of uploads lasting less than 33 seconds.

Perhaps the most telling shift in user behavior is video length. While early streaming platforms focused on long-form, TV-style content, the typical YouTube video in 2024 is just 64 seconds long, with over one-third of uploads lasting less than 33 seconds. This data underscores the growing dominance of short-form content, a trend accelerated by TikTok, Instagram Reels, and YouTube Shorts. The instant gratification loop of bite-sized videos has changed how users engage with media, favoring rapid, continuous consumption over deep, extended viewing.

As streaming evolved, user behavior became the driving force behind industry innovation. While platforms like Broadcast.com and early YouTube sought to expand access, modern streaming services have pivoted toward retention and monetization strategies, fine-tuning algorithms to keep users engaged for longer periods. The trade-off? More content than ever before, but far fewer opportunities for organic discovery.

This shift poses a critical question for the future of streaming: Does democratization of content still hold value if the majority of uploads remain unseen?

Early User Adoption and Resistance

In 2000, if you wanted to consume an audiobook on the go, you either needed to carry around a bunch of CDs (or, ugh… cassette tapes) and a portable player or an MP3 player. At Timberwolf Press, we mostly saw MP3 download sales going to college students. Of course, you couldn't fit a whole book on the MP3 player of the day – there wasn't enough storage space. However, we had some success with folks who downloaded an episode or two for their daily commute.

Netflix's transition from a DVD rental service to a streaming platform in 2007 marked a pivotal moment. Users had to rewire how they thought about access to content, from owning physical media to trusting that an online library would always be there. Hulu's launch in 2008 provided another bridge, easing hesitant users into the streaming era by offering next-day TV episodes alongside a growing library of past seasons.

By 2010, as broadband penetration improved and mobile devices became more powerful, streaming went from a niche option to the default way people consumed content. BBC iPlayer exploded in popularity, with millions of streams in just weeks. The same pattern repeated worldwide. Once audiences got a taste of instant access, traditional formats looked outdated.

How User Feedback Shaped Streaming Platforms

If the early 2000s were about proving that streaming was viable, the 2010s were about refining the experience. Consumers weren't passive. They demanded better quality, smarter recommendations, and more control over their viewing experience.

- **Better Streaming Quality:** Audiences complained about the pixelated video, buffering wheels, and lag, so platforms optimized their CDNs (Content Delivery Networks), codecs, and compression algorithms to deliver high-quality video with less bandwidth. This paved the way for HD streaming and, later, 4K and HDR content.
- **Smarter Recommendations**: Users wanted personalized content discovery, not just a library dump. Netflix's algorithm investments changed the game, analyzing viewing habits to predict what people wanted before they even knew it. While this arguably worked well for movies and shows (at least for some people, but never in any way that made sense for me, personally), music streaming services like Spotify nailed it even better. Their Discover Weekly playlists felt eerily spot-on, setting the gold standard for algorithm-driven content curation.
- **Evolving Business Models**: Initially, subscription models seemed like the only way to go. But as more platforms emerged, audience fatigue set in. People didn't want to pay for a dozen different services. This led to the rise of hybrid models like:
    - Ad-supported free tiers (Hulu, Peacock, YouTube's free movies)

o   Pay-per-view or event-based purchases

o   Bundled deals with internet and telecom providers

The Double-Edged Sword of Personalization: Data, Privacy, and Control

As streaming services refined their recommendation engines, they began collecting more data than ever before. At first, this seemed like a win-win. Users got better content suggestions, and platforms increased engagement. But soon, the sheer amount of personal data being collected raised red flags.

How Data Collection Works in Streaming

Every time you watch something, search for a show or even **pause a video**, platforms collect data to build a detailed profile of your interests. Some of the most commonly tracked data include:

- **Viewing History:** What you watch, how long you watch, when you stop watching.
- **Search Queries:** What you're actively looking for.
- **Interaction Data:** Likes, shares, reviews, even how long you hover over a thumbnail before clicking.

This level of deep tracking allowed companies to fine-tune recommendations and sell highly targeted ads. But it also led to growing concerns about privacy, data security, and content manipulation.

User Backlash and Privacy Concerns

By the mid-2010s, streaming platforms and social media companies were under fire for how they handled user data. Some of the biggest concerns included:

- **Data Overreach**: Platforms were collecting far more information than users realized.

- **Indefinite Data Retention:** Even after users deleted accounts, some platforms **kept their data indefinitely**.
- **Third-Party Data Sharing:** Many companies sold user data to advertisers and third-party firms without clear consent.

A 2024 Federal Trade Commission report called out major streaming and social media companies for lacking adequate privacy controls, particularly for younger users. In response, companies were forced to rethink their data practices, but only after years of quietly collecting and monetizing personal data.

The Future of Audience Engagement: Where Are We Headed?

Streaming services are now at a crossroads. Audiences expect more personalization, but they also demand more privacy controls. The challenge for platforms moving forward is balancing these competing priorities.

Some of the biggest trends shaping the next phase of user engagement include:

- More Transparency on Data Collection – Companies are being pushed to show users exactly what data is collected and how it's used.
- Stronger User Controls – Platforms are introducing opt-outs, privacy toggles, and auto-delete features to build trust.
- AI-Driven Content Curation – Expect more advanced recommendation engines and greater user control over what gets suggested.

Of course, it doesn't help that there is a common perception that our phones and digital assistances are constantly listening to us, whether we want them to or not, leading to ads and content subsequently showing up related to our presumably private conversations. Indeed, 27% of Americans don't use voice assistants because of privacy concerns. https://

Privacy concerns aside, the evolution of streaming has been driven by audience demands at every step. Better quality, smarter curation, fairer pricing, and stronger privacy protections. The platforms that can navigate these shifts while keeping users engaged will define the future of streaming media.

Innovation vs. Regulation: The Evolving Regulatory Landscape of Streaming Platforms

In the 1990s, the internet was pretty much unregulated. It was truly the Wild West. Rights contracts had no mention of the Internet. Applicable law meant stretching laws intended for other media to apply to the net. We were always aware that this could change suddenly in response to unforeseen issues. At Broadcast.com, we did our best to adhere to broadcast industry standards for our content, even though there were no formal rules governing streaming. To help navigate the uncertain legal landscape, we hired a Washington lobbyist to advocate for minimal regulations and ensure that streaming's growth wasn't stifled by reactionary policies. Since then, the regulatory environment for streaming platforms has evolved significantly, shaping how modern services operate today.

The 1990s: A Nascent Regulatory Framework

In the 1990s, the internet's rapid expansion outpaced the development of comprehensive regulatory policies. Early streaming services operated in a largely unregulated environment, focusing on technological innovation without significant legal constraints. The primary regulatory concerns during this period centered on issues like copyright infringement and the unauthorized distribution of content, as exemplified by legal battles involving platforms like Napster.

The 2000s to Present: Increasing Regulatory Scrutiny

In the 2000s, I was repeatedly approached by entrepreneurs eager to build their streaming services. Many of them believed that because certain content had aired on TV, it was fair game to rebroadcast online. Or they had some similar misconception, convinced they had found a loophole in copyright law. Let's just say they weren't particularly thrilled when I tried, as gently as possible, to introduce them to reality.

As streaming platforms grew in popularity and influence, regulatory bodies began to implement more structured policies to address various concerns:

- **Content Regulation and Local Content Requirements**: Governments worldwide have started to impose regulations on streaming services to ensure the promotion of local content and cultural diversity. For instance, in Canada, there have been discussions about amending the Broadcasting Act to include online streaming services, requiring them to contribute to the production of Canadian content. https://policyreview.info/articles/news/regulating-netflix-cross-national-momentum/1706

- **Taxation and Fees**: Several states and municipalities are attempting to extend franchise fees, traditionally levied on cable and internet service providers, to streaming video companies. This move aims to level the playing field between traditional broadcasters and streaming services. https://digitalcontentnext.org/blog/2022/10/27/policy-issues-on-the-streaming-video-horizon

- **Intellectual Property and Copyright Enforcement**: The rise of streaming has led to complex intellectual property challenges. Regulatory bodies are focusing on ensuring that streaming platforms respect copyright laws and fairly compensate content creators.

- **Consumer Protection and Data Privacy**: With the vast amount of user data collected by streaming platforms, regulations like the General Data Protection Regulation (GDPR) in Europe have been enacted to protect consumer privacy and govern data usage.
- **Net Neutrality and Access**: Regulatory debates around net neutrality impact how streaming services deliver content. Policies ensuring that internet service providers treat all data equally affect the accessibility and quality of streaming services.

Impact on Innovation

The evolving regulatory landscape presents both challenges and opportunities for streaming platforms:

- **Compliance Costs**: Adhering to diverse regulatory requirements across different jurisdictions can increase operational costs for streaming services.
- **Content Strategies**: Regulations mandating local content production influence the programming and investment decisions of platforms, encouraging the creation of region-specific content.
- **Market Entry Barriers**: Stringent regulations may pose challenges for new entrants, potentially stifling innovation by favoring established players who can navigate complex legal frameworks.
- **Consumer Trust**: Robust regulatory frameworks can enhance consumer trust by ensuring data privacy and content standards, potentially increasing user engagement and loyalty.

In conclusion, while the regulatory environment has become more complex since the 1990s, the stated goals often aim to balance the promotion of innovation with the protection of consumers and fair market practices. Streaming platforms must navigate this evolving landscape carefully, adapting their strategies to comply with regulations while continuing to innovate and meet consumer demands.

The Evolution of Live Sports Streaming: From Dial-Up to 5G

In the mid-to-late 1990s, one of the biggest drivers of Broadcast.com's rapid growth wasn't just music, business webcasts, or talk radio. **It was live sports**.

Back then, sports leagues and broadcasters were only just beginning to realize the power of the internet. The idea of streaming a game online seemed absurd to traditional media executives, many of whom still saw the internet as a fad. But sports fans saw the potential immediately.

We started small, streaming **SMU vs. Arkansas in 1995**, which became the first live college football game broadcast over the internet. The numbers weren't earth-shattering. At first. But the idea of a fan in Dallas being able to listen to a Razorbacks game from another state, in real-time, with nothing more than a PC and an internet connection? That was revolutionary.

The Early Challenges: Streaming Sports in the 90s

Sports were uniquely challenging to stream. Unlike music or pre-recorded content, live games demanded real-time delivery. But the internet of the 90s wasn't built for that. We had to navigate:

- **Slow dial-up connections** that turned a live broadcast into a buffering nightmare.
- **Limited bandwidth** forced us to optimize every bit of audio and video.
- **Lack of monetization models**, as traditional advertisers didn't understand the value of online streaming yet.

Despite these obstacles, sports fans adapted. By 1999, Broadcast.com was streaming more than 420 college and professional teams, including Super Bowls, the World Series, the NCAA Final Four, and even the Masters and U.S. Open golf tournaments. That success helped prove that sports weren't just another category of online content. They were the key to mass adoption of streaming media.

From Passive Viewing to Fan-Controlled Sports

At the time, sports streaming was still one-way communication. Fans could listen or watch, but they had no control over the experience.

- They couldn't switch between multiple commentators.
- They couldn't access real-time stats alongside live audio.
- They were locked into whatever was being broadcast at the moment.

Fast forward to today, and the landscape has completely changed. The demand for personalized sports content has never been higher. Fans now expect:

- Multiple commentary options, from official broadcasters to alternative feeds (e.g., ManningCast).
- Real-time stats, betting integrations, and instant replays embedded into their stream.
- The ability to choose how they engage with live sports, whether it's on their phone, smart TV, or in the stadium itself.

The Battle for Live Sports Streaming: Then vs. Now

The early 2000s saw a shift in who controlled sports streaming rights. While Broadcast.com was able to negotiate directly with teams and leagues in the 90s, today's environment is far more restrictive:

- Major platforms like Amazon, YouTube, and Apple are now competing for billion-dollar exclusive rights.
- Regional sports networks (RSNs) are struggling to adapt to

streaming-era distribution.
- Sports leagues are increasingly cutting out middlemen, launching their own direct-to-consumer streaming services.

The irony? The same battle we fought to make sports accessible in the 90s is happening again – just on a bigger stage.

A Significant but Overlooked Opportunity in Sports Streaming: Real-Time, In-Venue Audio

While the sports industry chases the next big thing – AR overlays, immersive fan experiences, and AI-driven analytics – a simple but powerful opportunity remains largely ignored: real-time, in-venue audio. Modern stadiums are filled with screens and digital enhancements, but fans still struggle to hear live commentary without lag. Unlike the high-bandwidth, high-latency solutions dominating the conversation, this opportunity is about delivering a thin stream, ultra-low-latency audio experience that enhances the game without distraction. It's not about reinventing the fan experience – it's about solving a problem that should have been fixed long ago.

If the '90s were about proving that live sports streaming was possible, today, a challenge I am focused on is solving an overlooked problem: listening to color sports commentary, live, in-stadium. While local AM and FM radio stations often provide this commentary, those signals don't propagate well in modern stadiums, leading to frustrating dead zones. Fans may try streaming the play-by-play on their phones, but no one wants to hear the call thirty seconds – or often much more – after the touchdown, goal, or home run has actually happened. These days, everyone knows that streams have too much lag to be practical for live, in-stadium listening.

That's where SportsBug comes in.

Unlike traditional streaming, SportsBug delivers live, in-stadium audio within 300 milliseconds, meaning fans hear the call as it happens – without delays, buffering, or dependence on expensive stadium

WiFi. Ironically, this challenge isn't new. The original SportsBug – an AM, and later FM, radio headset – faced similar issues, as those same stadium dead spots contributed to its decline. Now, we're bringing that idea into the modern era, solving the same problem with today's technology.

And just like how Broadcast.com made sports accessible beyond local radio stations, SportsBug is giving fans control over their live experience – right from their seats.

It's a really fun project that takes me back to my AudioNet roots. Instead of streaming the color commentary to alumni across the country, we're reversing the focus – bringing it full circle back to the fans in the stadium. Fun stuff!

The Next Wave of Innovation in Digital Media: AI, AR, and the Evolution of Content Engagement

The pioneering spirit of Broadcast.com in the 1990s laid the groundwork for today's dynamic digital media landscape. As we look ahead, artificial intelligence (AI), augmented reality (AR), and other rapidly advancing technologies are set to reshape how content is created, distributed, and monetized, echoing the same innovative drive that fueled Broadcast.com's inception.

Artificial Intelligence in Content Creation and Distribution

AI is transforming media in ways that would have been shocking just a few years ago. Generative AI models are assisting in producing music, video, art, and even real-time sports commentary, while advanced machine learning algorithms personalize content recommendations, keeping audiences more engaged than ever. AI-driven automated editing tools are streamlining production workflows, allowing creators to focus on storytelling rather than technical hurdles. Additionally, AI-powered real-time translation and dubbing are breaking down language barriers, enabling global audiences to access content in ways never before possible.

Beyond creation, AI is optimizing content delivery and analyzing user behavior to ensure relevant, engaging, and highly targeted experiences – a step beyond the early streaming personalization efforts pioneered by Broadcast.com.

Augmented Reality and the Future of Live Events

While much of the focus on media innovation centers around streaming, AR is still in its early adoption phase but is already beginning to transform live experiences – from interactive sports broadcasts to enhanced in-stadium engagement. Unlike traditional streaming, where engagement is passive, AR introduces a layered, interactive experience, allowing fans to see real-time stats, instant replays, and live player analytics overlaid onto the real-world environment. Whether through AR glasses, smartphone apps, or stadium screens, this emerging technology has the potential to reshape how audiences experience live sports, concerts, and entertainment as adoption grows and capabilities expand.

The Evolution of Monetization and Fan Engagement

Emerging monetization models are reshaping the digital content landscape by prioritizing direct creator-to-fan engagement. Platforms like **OnlyFans** and **Fansly** exemplify this shift, enabling creators to offer subscription-based communities where fans pay recurring fees for exclusive content and personalized interactions. These platforms also incorporate features such as direct messaging monetization, allowing creators to charge for private content or live sessions, thereby fostering a more intimate and profitable connection with their audience. https://ecommercefastlane.com/it/fansly-vs-onlyfans

In the gaming industry, **microtransactions** have become a prevalent monetization strategy. Mobile games often implement in-app purchases, allowing players to buy virtual goods, enhancements, or access additional content, thereby generating revenue beyond the initial game download. https://www.moengage.com/blog/mobile-gaming-monetization

AI-curated fan experiences are also gaining traction, with platforms utilizing artificial intelligence to deliver personalized content recommendations. For instance, AI-driven feed searches analyze user behavior to present tailored content, enhancing user engagement and loyalty. https://arena.im/audience-engagement/ai-powered-feed-search-for-fan-loyalty

The rise of **personalized content hubs** is evident in platforms like **Roblox**, which plans to enable creators and brands to sell physical merchandise directly from their virtual experiences. This approach allows users to interact with their favorite creators in immersive 3D spaces, fostering a more engaging and monetizable environment. https://www.voguebusiness.com/story/technology/as-gen-z-matures-virtual-stores-get-an-ai-boost

Interactive sponsorships are being utilized in innovative projects like **Rival Peak**, an interactive media experience that combines AI-driven simulations with real-time audience participation. This format allowed for the dynamic integration of sponsors into the content, creating a more engaging and interactive advertising experience. https://en.wikipedia.org/wiki/Rival_Peak

In the realm of **AI-driven merchandising**, advancements in artificial intelligence are enabling the creation of virtual stores that offer personalized shopping experiences. For example, platforms are developing AI-generated 3D content and virtual try-ons, allowing consumers to visualize products in a personalized manner before making a purchase. https://www.voguebusiness.com/story/technology/as-gen-z-matures-virtual-stores-get-an-ai-boost

Finally, implementing **real-time audience feedback loops** has become essential for content creators aiming to refine their offerings. Utilizing AI-powered tools, creators can gather and analyze user feedback promptly, enabling them to adjust content strategies in response to audience preferences and enhance overall engagement. https://martech.org/how-to-build-fast-effective-feedback-loops-in-an-ai-driven-world/

Collectively, these models signify a shift towards more interactive, personalized, and direct engagement between creators and their audiences, moving away from traditional, ad-heavy approaches.

Conclusion: Building on the Legacy of Streaming Pioneers

The early days of internet broadcasting were about proving that digital content delivery could break free from traditional media constraints. Today, that innovation continues – not through speculative decentralization but through practical advancements in AI, AR, and interactive engagement. While flashy, futuristic ideas often grab headlines, the real impact will come from technologies that solve existing problems, enhance user experiences, and create sustainable opportunities for content creators and audiences alike.

Just as Broadcast.com once redefined media consumption, the next wave of streaming evolution won't be about reinventing the wheel – it'll be about refining it.

The Rise of Streaming Platforms and the Decline of Traditional Cable News

In recent years, the media landscape has undergone a significant transformation, with streaming platforms experiencing substantial growth while traditional cable news networks face declining viewership. This shift reflects changing consumer preferences and technological advancements that have reshaped how audiences consume news and entertainment.

Decline in Cable News Viewership

Traditional cable news networks have seen notable decreases in their audience numbers. For instance, CNN's average primetime viewership declined by 25% from 1.1 million in 2021 to 828,000 in 2022. https://www.pewresearch.org/journalism/fact-sheet/cable-news/

This downward trend continued, with reports indicating that CNN's ratings reached historic lows in subsequent years. Headline: "CNN Faces 'Unprecedented Crisis' After Historic Ratings Collapse: 'Only 61,000 Viewers in Key Demo'"https://www.econotimes.com/CNN-Faces-Unprecedented-Crisis-After-Historic-Ratings-Collapse-Only-61000-Viewers-in-Key-Demo-1694127

In response to these challenges, CNN announced plans to lay off approximately 200 employees, representing about 6% of its workforce, as part of a strategic shift towards digital platforms. The network is investing $70 million into its digital business and developing a new streaming service to adapt to changing viewer habits. https://www.ft.com/content/e1368295-b24e-4295-8ff8-7015714286fe

Growth of Streaming Platforms

Conversely, streaming platforms have experienced significant growth. As of 2024, 64% of U.S. viewers subscribed to both traditional TV and video streaming services, while 33% relied exclusively on streaming services. https://www.statista.com/statistics/1385919/traditional-tv-streaming-viewers-us/

Americans now spend an average of three hours and nine minutes per day streaming digital media, highlighting the central role these platforms play in daily entertainment consumption. https://www.forbes.com/home-improvement/internet/streaming-stats/

The flexibility and convenience offered by streaming services, such as on-demand content and personalized viewing experiences, have attracted audiences away from traditional cable TV. Additionally, the rise of ad-supported streaming platforms like Tubi and FreeVee provides cost-effective alternatives for viewers, further contributing to the decline in cable news viewership. https://blog.gwi.com/marketing/streaming-trends-and-stats/

<u>Implications for the Media Landscape</u>

The contrasting trajectories of streaming platforms and traditional cable news networks underscore a broader shift in media consumption habits. As audiences increasingly favor the flexibility and diverse content offerings of streaming services, traditional media outlets are compelled to innovate and adapt to maintain relevance in a rapidly evolving digital environment.

Scaling Up: The Hidden Complexity of Streaming Infrastructure

As streaming grew from an experiment into a viable industry, the challenge was no longer just proving it could work – it was figuring out how to handle the sheer volume of content that needed to be ingested, processed, and delivered. What started with a handful of faux live streams in 1995 quickly expanded in the following months and years into thousands of simultaneous broadcasts, each requiring different formats, connections, and logistical solutions.

This wasn't just a software problem – it was a massive operational challenge. Content didn't just appear online ready to stream; it had to be captured, encoded, stored, and distributed, often with custom-built systems because nothing off-the-shelf could handle the scale. From receiving feeds via satellite and telco lines to managing an ever-growing library of recorded media, every aspect of the process required constant adaptation.

And sometimes, it wasn't just technical hurdles that needed solving – there were geopolitical and regulatory challenges, too. That's where Broadcast.com's global expansion took streaming into uncharted territory, leading to situations that were as challenging as they were memorable.

To carry all the content, we had to ingest it. This came in many forms. Recorded content came in on various media we had to encode. Almost nothing was already encoded. Live content came in via telco and satellite. Some programs sent us a dish to receive their feed. Our roof was crowded with a huge array of over twenty different satellite dishes in what I was told was one of, if not, the largest satellite farms in the southwest.

International Incident

I clearly recall receiving a jarring phone call from my Intel contact in China at around 3 a.m., urgently requesting my presence in Beijing within just three days. At first, I wasn't even sure if it was physically possible. Merely obtaining my visa required frantic coordination – rushing my passport down to the Chinese Consulate in Houston for processing, anxiously awaiting the visa stamp, and then racing to have it flown back in time. As soon as the passport returned, I was on a flight to Tokyo and then immediately on to Beijing, barely making the tight timeline.

Landing in Beijing late at night felt surreal, almost cinematic. Gazing out of the airplane window, I remember vividly seeing welding torches scattered across the landscape, their sparks glowing through the heavy smog, giving the entire city an eerie Blade Runner-like appearance. Nixon's historic journey to China crossed my mind – another American venturing into what was still largely uncharted territory for most Westerners at the time.

Stepping off the plane, I was greeted by the imposing sight of jackbooted PLA soldiers at the airport, a stark reminder that less than a decade had passed since the tragic events of Tiananmen Square. Though everyone I encountered was impeccably polite, it was impossible to shake the strange and disorienting feeling. Standing there, this Texas boy in his boots, navigating an unfamiliar landscape steeped in historical significance, was an experience unlike anything I'd ever imagined.

The next morning, after staying at the Shangri-La Hotel, which became my usual spot, I vividly recall having to persuade a huge boardroom of government officials that their idea of routing our broadcast signals through Shanghai, then to Beijing, and finally out of the country simply wasn't feasible. Beyond the technical nightmare, I gently suggested it wasn't in their best interest anyway. Somehow, standing there – a Texas boy in a suit and boots, scribbling diagrams on a whiteboard with the assistance of a UN translator – I managed to convince them. I was so much in the zone that I don't clearly recall the presentation beyond the intensity of it all. Frankly, even I was surprised it worked.

Andy Grove

Andy Grove, the legendary Chairman of Intel at the time, frequently headlined many of the live events we broadcast – including several high-profile presentations in China. The dynamic relationship between China and Intel was particularly intriguing, as Grove enjoyed an almost iconic status there, widely revered as the "father of the PC." His influence was remarkably visible: bicycles were adorned with Intel stickers, entire truckloads of computers and components would mysteriously arrive, and families would pool their savings together just to purchase a single PC.

Our responsibilities during these landmark events were significant and complex. To ensure that the webcasts could reach audiences regardless of their connection speed, each live broadcast was meticulously encoded at multiple bitrates. Typically, we started as low as 14.4 kbps, stepping incrementally upward through 28.8 kbps and 56 kbps, and often extending as high as around 300 kbps. This range of streams allowed us to accommodate virtually every level of internet connectivity, from dial-up users patiently navigating early modems to viewers enjoying the relative luxury of faster DSL or corporate T1 connections.

For every bitrate stream, we operated three dedicated PC encoders – a primary encoder and two backup units, ensuring absolute redundancy. The backups weren't just there for appearance. The encoding software at that time was flaky. We *needed* those backups because we couldn't just reboot a stream – we had to switch to the backup running in parallel to avoid knocking stream viewers off the streams, killing the broadcast.

Alongside these encoding PCs, additional computers managed content and slide presentations, all integrated seamlessly with our extensive array of audio and video equipment.

Backstage, our setup resembled a sophisticated broadcast command center rather than a typical webcast workstation. With numerous PCs, monitors, switchers, mixers, and racks of gear meticulously arranged, the visual impression alone was impressive, underscoring both the complexity and the scale of these groundbreaking live events. Each event was not just a technical challenge but a testament to the rapidly evolving landscape of digital communications, showcasing Intel's vision and Broadcast.com's pioneering capability in real-time global broadcasting.

The late great former Intel Chairman Andy Grove once commented on our equipment we set up for his event in China, saying, "You could launch a moon rocket from here!"

The Spy

During one live event trip, I found myself with a rare few extra hours in Shanghai and decided to take advantage of the opportunity to stretch my legs and explore the city beyond the confines of my hotel. The streets were alive with activity – bustling crowds, vibrant street vendors, exotic aromas, and an overwhelming array of sounds that were fascinatingly different from my routine life back in Texas. Normally, I make an effort to experience local cuisine and immerse myself in the culture of the places I visit. However, this time, I found myself pressed for time and, admittedly, feeling somewhat cautious. After wandering for a while and not immediately spotting another eatery that appeared, let's just say, reassuringly approachable, I came across an unexpected sight: a McDonald's. I'm not a huge McDonald's fan, but it seemed like a safe enough option.

The menu was entirely in Chinese, but the presence of familiar pictures provided sufficient reassurance for my tired and hungry self. I pointed to something that appeared reliably familiar and confidently paid with my trusty American Express card. Finding an empty table, I settled in to enjoy my meal, grateful for this brief moment of familiarity amidst a day of cultural adventures.

That's when the surreal became truly cinematic. Approaching my table was an impeccably groomed gentleman with neatly greying hair, dressed in an elegant three-piece suit complete with a bow tie that suggested sophistication and formality. He politely introduced himself, explaining that he simply wished to practice his English, though his crisp Oxford accent made it abundantly clear that he wasn't really there for that reason. He was there to question me, specifically.

Our initially friendly exchange soon transitioned into something far more deliberate and calculated. With polished courtesy and practiced ease, my impeccably dressed companion gradually guided the conversation toward more pointed territory. He began to ask increasingly detailed questions about my frequent travels to China: precisely whom I was meeting, the specific purpose of each visit, and my broader objectives regarding our activities in the region. It quickly became clear that this wasn't merely an innocent encounter designed to practice language skills – it was an interrogation, expertly concealed beneath layers of polite conversation.

Suddenly, my casual fast-food lunch felt significantly less relaxed. With heightened awareness, I carefully measured each response, acutely conscious that my answers had potential implications not just for me personally but also for my company and our critical partnership with Intel.

My answers seemed to satisfy him. I don't know if it was my imagination; however, my fellow team members and I felt like we weren't being followed quite as much after that.

Lobster, VIPs, and Rotating Antennae

There was one particularly memorable dinner during my travels in China – one of those formal hosting occasions where the goal was clearly to impress the visiting foreigner. My hosts took me to a restaurant that, judging from the architecture alone, had a very clear specialty. How could I tell? Because the entrance was encased in a giant clear Lexan archway – essentially a walk-through aquarium – filled with live lobsters. You literally entered the restaurant by walking under and through a tunnel of lobsters. Message received.

With great formality, they escorted me through the restaurant, proudly showing off their VIP guest to staff and diners alike before ushering us into a private dining room. At the center was an enormous round table fitted with a giant lazy Susan. The meal was to be served in courses, a progressive feast of regional specialties.

Now, as is customary, the meal opened with a sort of... initiation course. A traditional delicacy – something I couldn't quite name and, frankly, prefer not to remember. As the guest of honor, I had the privilege of tasting it first. I don't drink, but in this situation, alcohol was very much required. I managed to wash it down with a generous pour of baijiu, suppressing both the flavor and any outward signs of hesitation.

The walls were lined with identically dressed serving women, moving in practiced unison as they brought out dish after dish. I don't mean to be rude or insensitive or whatever, but to my inexperienced Texas boy eye, the women could have been clones. I'm sure they were picked because they were the same height, etc., and once dressed and made up to look alike, well, they did. Anyway, once the meal got rolling, I have to say that 99% of the food was absolutely fantastic. Incredible flavors, textures, and craftsmanship.

Then came the salmon.

Rather, the salmon was served on a lobster.

What appeared was the bottom half of a very large lobster, spread open like a serving dish, with beautiful slices of salmon draped across it. The rest of the lobster – the head, claws, and massive antennae – remained intact. Only those antennae were... moving. Slowly sweeping in wide, almost deliberate arcs.

You could feel the shift in the room. Several guests turned noticeably green. Quiet, urgent whispers were exchanged with the serving staff. A few moments later, someone entered the room with a pair of kitchen shears and – without ceremony – snipped off the antennae. Unfortunately, this didn't entirely solve the problem. The stubs left behind continued their ghostly movement. Two or three of the women excused themselves, returning later looking visibly paler.

The rest of us soldiered on. I did what any good guest would do: smiled politely, nodded appreciatively, and took very small, respectful bites.

Fried Rice

On one of our trips to China, Ed Luczycki and I flew in after a long haul aboard China Air. I'll just say the in-flight meal experience left a lot to be desired. By the time we landed, cleared customs, and made it to the hotel, we were completely wiped out – tired, hungry, and not in the mood to explore the local culinary scene too deeply.

Thankfully, Intel had booked us into the Four Seasons or something very close to it. At the time, Intel had a policy that required vendors like us to stay at the same hotels and fly the same class as their own people. It was a perk we didn't complain about – and we made the most of it while it lasted.

So Ed and I wandered into the hotel restaurant, hoping for something quick and comforting. The menu was full of elaborate dishes, most of which required more brainpower than we could muster at that point. Then we spotted something simple: fried rice. Reliable. Familiar. Exactly what we needed. We both looked at each other, nodded, and told the waiter, "We'll each have the fried rice."

The waiter smiled and replied with enthusiastic clarity: "Yes! One fried rice."

Ed and I exchanged glances. "No," we said, "we each want one." He nodded again, "Yes. One." At this point, the back-and-forth started to resemble a Monty Python sketch. We tried to explain – two people, two orders. Still, the waiter looked confused. He gestured politely, trying to convey that it was meant to be shared.

Now, I'm six feet tall. Ed is at least two inches taller than me – maybe four, depending on the boots. We're not small guys. And after a 14-hour flight and next to no food, the idea of sharing a plate of fried rice wasn't exactly our vision of dinner. But we relented.

The fried rice came. It was good. Very good, actually. And it disappeared almost instantly. Still hungry, we flagged the waiter down and asked for another order.

That's when things took a turn. The waiter turned pale – visibly alarmed. He lost several shades of color and looked genuinely concerned. Not annoyed, not confused. Just… afraid. He gave a stiff nod, vanished into the kitchen, and we never saw him again. We returned to that restaurant several times over the next few days, but he was simply gone.

To this day, we've wondered whether he got in trouble for letting the giant Americans double down on the fried rice. Maybe we broke some unspoken culinary protocol. Maybe he was reassigned to a remote hotel in Inner Mongolia. We'll never know. But we still feel a little guilty about it.

On another occasion, at another establishment, Keith Brown tried to order Terriaki Chicken. He got a huge pan full of hot water and a chicken floating in it...

China Infrastructure Challenges of the 1990s

The logistics of doing a live event in China in the 1990s were monumental. Before every broadcast, our team traveled to the venue weeks in advance to meticulously plan and inspect every detail. We arranged for specialized power conditioning and conversion equipment (because the power available there was poor quality and would induce problems in broadcast equipment) and set up multiply redundant telecom connections. At the time, their telco was... obsolete, even by 1990's standards.

Logistics for these events was complex. With all the computers and gear, we needed to ship 12,000 pounds of tech. Just the shipping and customs issues were problematic enough. Also, we often had more than one event going on, so we'd need to have more gear available for those events, too.

The Defector

We'd meticulously prepared for another major live event in China for Intel, undertaking two detailed advance trips to ensure everything was perfectly arranged, from power conditioning to telecommunications redundancy. When we arrived the third time, confident and ready for the live broadcast, the entire situation quickly deteriorated into chaos. Almost immediately, we realized that the carefully specified power conditioning and conversion equipment was completely wrong – worse than inadequate, it was outright disastrous. Details we'd painstakingly arranged were either entirely missing or disturbingly incorrect.

As the broadcast began, the true nightmare unfolded. All nine redundant telco lines, our carefully planned safety net, simultaneously began cycling on and off without explanation. It wasn't a technical malfunction – it was clear sabotage, deliberate interference that crippled our broadcast. Desperate attempts to stabilize the situation proved futile as the lines continued their relentless, erratic disruptions.

The atmosphere was tense, and confusion quickly turned into intense frustration. This wasn't just any broadcast; it was for Intel, one of our largest and most critical clients, whose fees quite literally kept our business afloat. I'd never before faced such a complete and inexplicable failure on a broadcast. The magnitude of the moment hit me hard, and in a state of genuine anguish, I turned to Intel's Peter Morgan, feeling utterly devastated, and said, "Sometimes the space shuttle crashes, Peter. I'm sorry."

It wasn't until later that we discovered the unsettling truth: the local China Telecom employee who had been instrumental in arranging our telecommunications had defected to Taiwan. Apparently, powerful authorities were expressing their anger and frustration at his defection through deliberate sabotage of our event.

Thankfully, despite the catastrophic technical breakdown, Intel remained steadfastly supportive. Behind the scenes, Peter Morgan and his colleagues managed the fallout, and somehow, miraculously, we didn't lose the event fees. I'll likely never know exactly what happened in those closed-door discussions, but Intel's loyalty and support in that moment of crisis left an indelible mark on me, turning a nightmare scenario into a testament of partnership and trust.

Closing Summary for Chapter 6

The evolution of streaming has been a story of relentless innovation and shifting audience expectations. From the early days of Broadcast. com pioneering live sports streaming to today's battle for exclusive digital rights, one thing has remained constant – the demand for access, personalization, and real-time engagement.

The 1990s streaming revolution was about breaking free from the constraints of traditional broadcasting, proving that live content could be delivered online. The 2010s refined this model, with platforms focusing on better video quality, smarter recommendations, and new monetization strategies. Now, in the latter half of the 2020s, we are at the next major turning point: fans don't just want to stream games. They want to control how they experience them.

The story of streaming is still being written. The platforms that embrace real-time engagement, fan customization, and seamless experiences will define the future. The next big shift is already happening – will you be part of it?

Streaming transformed sports once. It's about to do it again.

- How has streaming changed the way you experience live sports?
- Do you still watch games on traditional TV, or have you switched to streaming platforms?
- Would you listen to in-stadium commentary if it was real-time, with no delay?

Stay tuned – Chapter 7 dives into the fierce competition for media rights and what it means for the future of fan engagement!

7: Broadcast.com's Streaming Odyssey

From CDNs to Yahoo's Big Miss

In the early days of streaming, latency was a significant challenge, often resulting in delays and buffering that hindered the user experience. To address this, I set up servers around the USA and across the globe, effectively creating an Edge Network. This strategic distribution network allowed us to route listeners and viewers to servers closer to their geographical locations, thereby reducing latency and enhancing streaming performance. Distributing content out to this network of edge servers also reduced the number of streams coming out of our Dallas-based server farm, spreading the load. To my knowledge, this initiative marked the inception of the first Streaming Media Content Delivery Network (CDN).

Impact on Streaming Media

Implementing an Edge Network was a pivotal moment in the evolution of streaming media. By reducing latency and improving content delivery speeds, we were able to provide a more seamless and enjoyable experience for our audience. This innovation not only enhanced the quality of our streams but also set a precedent for future streaming platforms.

Approaching Live Events Like True Broadcasters: The Secret to Our Success

In the early days of AudioNet and later Broadcast.com, one of the most significant challenges we faced was the lack of robust streaming technology. The encoders and servers of the time were prone to (frequent) failure, often requiring manual rebooting to fix issues. These glitches would cause the stream to drop, forcing users to manually reconnect to continue listening or viewing.

At this point, most of our competitors treated these issues as simple IT problems. That is when the encoder or server glitched, they would reboot the system, hoping the issue would resolve itself. But we took a very different approach. We decided to treat our live events like professional broadcasters would. In traditional broadcasting, if an issue arose during a live event, it wasn't simply fixed with a reboot and forgotten. Broadcasters had redundant systems, failovers, and backup plans in place to ensure that the show went on WITHOUT INTERRUPTION, no matter what.

We implemented multiple levels of redundancy at every stage of our streaming process – from the servers that hosted the streams to the encoders that handled the media and even the staff managing the events. Beyond that, for high-profile events, we engineered multiple data pathways with multiple telco connections to deal with bandwidth problems.

I should point out that it was not unusual to have bandwidth problems. The networks simply weren't built for the kind of uptime people have grown used to in later years. This was especially true when we broadcast live events from remote locations. Whether it was the first live streaming webcast from inside China, or *The World Economic Forum* from Davos, or the Funeral of Princess Diana, or product launches from Microsoft and Intel from around the globe, getting redundant bandwidth in place in advance, as well as shipping and setting up the huge amount of equipment needed was a logistical.... Challenge.

We ensured there was always someone on hand to monitor the

streams, address issues proactively, and keep the audience engaged. Even if a server or encoder failed, our backup systems would kick in without interrupting the broadcast. In many cases, we had redundant event staff available to address problems immediately, just as you would see in a traditional broadcast environment.

This approach significantly set us apart from the competition. While others were treating streaming like a technical fix, we were treating it like an ongoing, live production. This commitment to excellence and to treating every event like it was broadcast-quality content paid off. Our reputation for running smooth, reliable live events grew rapidly, and for a long time, live events became our largest revenue source.

Our attention to detail and professionalism during live events created a sense of trust with our audience. They knew that when they tuned in, they would experience a high-quality broadcast, generally free from interruptions or technical hiccups. That trust was critical in helping us build a loyal user base, attract more advertisers, and eventually scale our operations to new heights.

We had regularly handled broadcasting duties for a major music festival on behalf of a significant client. However, one particular year, a competitor swooped in with an offer to do the job at a substantially lower price. My contact with the client was clearly uncomfortable breaking the news; I listened patiently, then simply shrugged and asked him to check back in afterward to let me know how things turned out.

The short version? From what I heard, it was a complete disaster. Among the numerous missteps, the most glaring was the competitor's failure to secure necessary releases from the performing artists – a fundamental requirement. Frankly, that oversight alone spoke volumes about their experience (or lack thereof). Soon afterward, the client called me back – lesson learned – and from that point forward, we resumed handling the broadcasts.

The following image is an actual diagram from my original files from one of the simple events we did, with only 56k and 28.8k encoders. I recall that it was faster and easier for me to create quick diagrams like this in paint rather than using a diagraming tool. Remember, this was 1999:

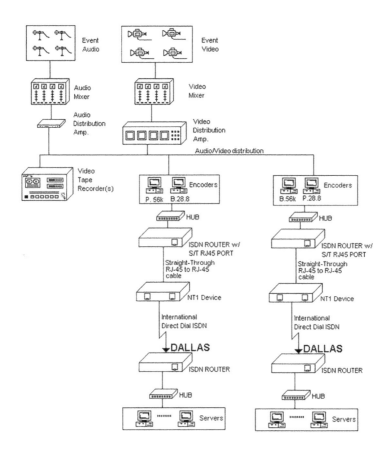

I firmly believe that our decision to treat live-streaming events like traditional broadcasting – with full redundancy and professionalism – was a key factor in our early success. It wasn't just about the technology; it was about delivering a consistent and professional experience to our users, which, in turn, helped establish Broadcast.com as a leader in the industry.

The Cost of Streaming in the 90s: A Key Challenge for AudioNet/ Broadcast.com

At AudioNet and Broadcast.com, one of the most significant obstacles we faced was the cost of bandwidth. The technology was new, the infrastructure was limited, and the costs associated with broadcasting live events over the Internet were hideously high. This made it one of the biggest challenges we had to overcome.

We broadcasted a mind-numbing array of live events. It was honestly a blur. In sports that included everything from college football games to the MLB World Series to the Soccer World Cup to the NFL Super Bowl, we had to handle all aspects of content delivery, from encoding to streaming over telecommunications and satellite links, as well as, in many cases, staff and equipment on-site at the live event. The high costs of bandwidth, combined with the technical limitations of the period, made delivering high-quality streams a financial burden.

By 1999, it was clear that the infrastructure we were building was extensive and costly:

- 700+ multimedia streaming servers
- 22 satellite receiving dishes
- Private frame relay network, as originally designed by Sherman Swartz
- Support for 650 simultaneous live events

At the time, we were dealing with multiple bandwidth connections in the 45 Mbps to 155 Mbps range, with numerous 45 Mbps and 155 Mbps connections (multi-homed). As we were broadcasting major sports events, we were often looking at an audience in the hundreds of thousands. It was enormous at the time but a far cry from today's standards of nearly instant, unlimited-scale streaming.

The costs associated with scaling these infrastructures were compounded by the fact that server uptime and bandwidth weren't as reliable as they are today. Encoders would frequently crash, and we were forced to manage both redundancy and failover systems just to ensure the stream remained live, requiring constant technical support.

Interestingly, I was recently reminded about how the Real Server end client limit was a number kept in a plain text config file, which we could then simply replace to match the network limit of the Dell server instead of the license restriction. And, of course, our servers were Dell Desktops sitting on top of shelves in racks (after the move to Taylor Street when we had an actual 'data center' room and not just machines piled on top of cafeteria tables until they were bowed down in the middle).

As I'll discuss in a moment, one of the big reasons for reliability problems was due to instability in the RealNetworks platform.

Evolving Our Codec Standards and Platform: A Strategic Shift in Technology

In 1995, the landscape of internet streaming was still in its infancy, and the technology available to support it was often cumbersome and expensive. Competing codecs were in the market, some requiring proprietary hardware that drove up costs. The largest internet sites relied on high-end UNIX servers, and to run those servers, you needed specialized UNIX administrators – a niche skill set that was costly and hard to find.

However, at AudioNet, I made a strategic decision that would have a lasting impact on our business: I standardized us on PCs running Windows and Windows NT instead of UNIX systems. This decision was driven by the need to cut costs due to our threadbare startup funding and to streamline operations. By choosing Intel-based PCs (later standardized on Dell Optiplex PCs), we could run Windows NT Servers, a much more affordable and user-friendly platform compared to the high-end UNIX systems that most other streaming services were using. Windows NT was point-and-click, which meant that we could hire less expensive techs to manage the servers and encoders rather than relying on highly specialized UNIX admins.

This decision not only made our operations more cost-effective but also opened the door to greater scalability and ease of maintenance.

There's a story there. We had been doing live Intel product release webcasts for a while. Eventually, word got up the food chain there and people started asking who the heck AudioNet was. My contact said that he'd like to bring some Intel executives out to see our shop.

Now, let me paint a clear picture for you. At that early stage, when we were still tightly packed into Mark's former bachelor pad at 2929 Elm Street, we were far from the polished image of a professional broadcasting facility. To put it mildly, we didn't exactly have server racks or neatly organized cable management. Instead, we had cafeteria tables loaded with so many Dell PCs that they visibly sagged under the weight, bowing alarmingly at their centers. (About once a week Todd and I would go and pick up another couple of cafeteria tables and haul them back to Elm Street).

Crickets

We also had a few machines tucked away at a local ISP called PicNet for a while. I still remember the smell. It hit you the moment you opened your car door. There were certain times of year – usually at night – when the entire parking lot and the sides of the building would be absolutely swarming with black crickets. Not a few. Millions. They blanketed the pavement and clung to the walls like something out of a horror movie. You couldn't walk without crunching them underfoot.

It wasn't just PicNet, either. Grocery store parking lots, strip malls – places all over North Texas seemed to get hit during those years. I don't recall it being that bad before or after, but maybe that's just how I remember it. Still, whenever I think of those late-night trips to PicNet to swap out a drive or reboot a server, that smell – part musty, part electric, part cricket – is one of the first things that comes back to me. You never forget it.

Elm Street

I keep getting off track. Elm Street. Oh my.

Ethernet cables hung haphazardly from the ceiling and spilled across floors, resembling dense jungle vines more than structured wiring. Nothing about our setup was fixed or permanent; nearly every PC – including, occasionally, even the Secretary's – could be quickly commandeered, reconfigured, and pressed into broadcasting service at a moment's notice. It was a controlled chaos – a perpetual state of improvisation that somehow, against all odds, consistently delivered groundbreaking broadcasts.

There were times, later when we were also using Windows Media, that we had to convert every machine to WM or Real overnight for some event or other, and then back again.

Oh, and 2929 Elm Street was an old, renovated building. There was black… stuff… on the ceiling. Whatever it was, it sort of gently rained down on us in specs over time. You never left your glass or cup uncovered unless you wanted to drink the stuff.

Intel's Visit to Elm Street

Getting back to the story here about Intel coming to see our place, needless to say, I couldn't exactly say… "no" to Intel. On the other hand, I figured as soon as they walked in the door, they'd probably drop us and run away screaming. So to speak.

Instead, the look of wonder on their faces was a shock. Their question? "You mean, you run EVERYTHING on Intel-based machines????"

The result? Long story shorter. Intel's involvement was so significant that they became our second capital investor (after Motorola), recognizing the potential of the Intel-based Windows NT platform for internet broadcasting.

More than that, Intel recognized that AudioNet was creating a market among consumers who would want faster PCs so they could consume the content that we were streaming. PCs with Intel Inside. Intel wanted to sell more chips.

Side Note:

Amidst our otherwise chaotic setup, we did have one makeshift equipment rack tucked away in the corner of the room – at least, I recall it as a rack of sorts, possibly double-sided. Standing next to it was a unique sensory experience; you'd be immersed in a layered symphony of multiple simultaneous live events, radio stations, and streaming feeds, each overlapping in a constant, rhythmic hum of activity. Remarkably, after spending enough time near it, you became so attuned to the familiar flow of audio streams that you could immediately detect if something went wrong. Even amid dozens of overlapping broadcasts, your ears learned to pinpoint precisely when a single feed faltered, glitched, or stopped entirely. It was a peculiar skill, borne purely out of necessity and daily exposure, underscoring the unusual yet dynamic nature of our early broadcasting environment.

Side, Side Note: Mark's House

As previously mentioned, on more than one occasion, I had to break into Mark's house, usually by climbing the broken, rotted wooden ladder to the second floor. He would leave his door unlocked, but his brother would stop by and lock it.

You see, Mark had an ISDN line installed at his house to support his LexisNexis setup, and that gave me yet another location where I could set up an encoder and broadcast stream. On more than one occasion, I'd find myself alone at his place, deep in technical troubleshooting, cables everywhere. I distinctly recall one afternoon, absorbed in working on the equipment, when the door swung open.

Suddenly, two very attractive young women walked confidently into the room, stopped abruptly in their tracks, and stared at me in wide-eyed confusion. After a moment of awkward silence, one of them finally managed to exclaim, clearly shocked, "You're not Mark!" I sheepishly agreed that, indeed, I wasn't. They quickly apologized and departed, leaving me both amused and mildly puzzled – but somehow not entirely surprised.

Strange Bedfellows: Microsoft & Broadcast.com

The strangest thing at the time, was the fact that Microsoft became a huge supporter. Before they began helping us, I had a constant worry that they could easily throw a ton of money at their own efforts and simply crush us. That they were ultimately going to be a huge competitor. Instead, they evidently saw us as a nimble startup they could leverage at minimal investment in order to make their Windows Media platform highly competitive with Real Media. Apparently, after Rob Glazer left Microsoft, there was no love lost, or at least, that was what folks like us thought.

Codec Evolution: From RealAudio to the Dual-Platform Strategy

When it came to selecting a codec, we initially relied on RealAudio – one of the first widely used codecs for streaming audio. RealAudio worked well for basic audio streaming, but it wasn't without its issues. The codec had been originally written for UNIX platforms, and when ported over to Windows NT, it suffered from consistent memory leaks that caused reliability problems. Despite these challenges, we continued to use RealAudio because, at the time, it was one of the only really viable options available for internet audio streaming at scale.

However, as we continued to grow our operations and attract more attention and press, RealNetworks began to compete directly with us for internet broadcasting business. Led by Rob Glaser, RealNetworks sought to capitalize on the growing demand for internet broadcasting. But despite their efforts, in the long run, RealAudio's codec wasn't able to match the stability and reliability that was needed for large-scale streaming.

Microsoft recognized the growing potential of Internet broadcasting. They were eager to support AudioNet in ways that RealAudio couldn't and introduced Windows Media Player and Windows Media Audio (WMA), offering a competing solution for internet streaming. Unlike

RealAudio, Microsoft's Windows Media codec was built to run natively on Windows NT, providing a much smoother experience and eliminating the memory leak issues we had faced.

As a result, we began to use both RealAudio and Microsoft's Windows Media codec for our broadcasts. Microsoft's active involvement was critical to our success. Not only did they provide the tools we needed to broadcast reliably, but they also provided significant and better technical support for their platform. Their active support gave us leverage against Real. BTW, I don't know when or if Real ever fixed their memory leak.

The Impact of Codec and Platform Evolution on Streaming

The evolution of our codec standards and platform played a central role in our ability to scale and deliver reliable content. By standardizing on Intel-based Dell PCs running Windows NT, we created a more cost-effective and scalable system than the high-end UNIX-based solutions used by most competitors. Furthermore, our decision to adopt both RealAudio and Windows Media as our streaming codecs ensured that we were able to meet the needs of our growing audience while avoiding the technical limitations of any single solution.

Ultimately, the combination of Intel's hardware, Microsoft's software, and RealNetworks' codec was instrumental in propelling us to new heights. These partnerships not only helped us solve our technical issues but also positioned us as the world's leading internet broadcaster, delivering high-quality, reliable streaming to an ever-expanding audience.

The Yahoo Acquisition & The Missed Opportunity

In 1999, Yahoo acquired Broadcast.com for $5.7 billion. It was one of the largest tech acquisitions of the dot-com boom. The vision was to turn Yahoo into a media giant, leveraging our streaming tech.

However, several factors contributed to the deal's failure:

- Lack of interest in the broadcast.com team: From my own observations and from insider sources, it was clear that Yahoo had almost no interest in leveraging the expertise provided by broadcast.com's existing team. **There was no synergy plan**.
- **Lack of Integration Strategy**: Yahoo struggled to integrate its numerous acquisitions effectively. Broadcast.com was no exception. Without a clear plan to merge technologies and cultures, the acquisition couldn't deliver the anticipated synergies. https://www.feedough.com/why-yahoo-failed
- **Shifting Focus and Identity Crisis**: Yahoo's indecision about its core business – oscillating between being a media company and a technology company – led to strategic disarray. This lack of focus hindered the effective utilization of Broadcast.com's assets. https://tribune.com.pk/story/1153035/yahoos-demise-internet-giants-failure-story-missed-opportunities
- **Rapid Technological Evolution**: The swift pace of technological advancements rendered some of Yahoo's acquisitions, including Broadcast.com, less relevant. Yahoo's inability to adapt these platforms to the changing tech landscape resulted in their obsolescence. (https://em360tech.com/tech-article/why-did-yahoo-fail

Over the years, I often heard it said that, "Yahoo is where startups go to die." It happened with broadcast.com, GeoCities Flickr, and even Tumblr.

Streaming's Evolution: What's Changed, What Hasn't

The journey from AudioNet to Broadcast.com was about more than just technology – it was about reimagining how live content could be delivered on the Internet. We pioneered real-time streaming, built one of the first CDNs, and approached live events with the redundancy and professionalism of traditional broadcasters. Our relentless focus on reliability and scale set the foundation for the streaming industry that thrives today.

Yet, despite nearly three decades of advancements – better compression, fiber networks, and cloud computing – some challenges remain. Latency still disrupts live sports, bandwidth costs remain high, and true real-time streaming remains an unsolved problem.

Streaming has transformed the way the world consumes content, but the mission isn't over. The next frontier isn't just about higher resolution or bigger libraries – it's about eliminating delays and creating a seamless, real-time experience for audiences everywhere.

What are your earliest memories of online streaming? Did you experience AudioNet or Broadcast.com in action? Let's talk about how far we've come – and where we're going next.

Why This Matters Today

Streaming has become the dominant way people consume content, but the underlying technology still hasn't fully solved the latency problem. In most cases, it doesn't matter – no one notices a delay when watching Netflix.

But in live sports, real-time betting, and interactive broadcasts, latency kills the experience. This is the challenge that still needs solving – one that we're working on today.

What are your earliest memories of online streaming? Back in the day, did you listen to AudioNet.com or watch broadcast.com programs and events? Did you watch YouTube in its early days? Let's talk about how far we've come – and where we're going.

8: Broadcast.com to YouTube

The Rise of User-Driven Media

When we launched AudioNet / Broadcast.com, we were laying tracks for a train that didn't yet exist. The internet was dial-up modem slow, video compression for online streaming didn't really exist yet, and streaming, as we know it today, was a concept far ahead of its time.

Back then, if someone wanted to stream content, they had to send us their audio and, in later years, their video, which we would encode from tape or via satellite and broadcast it from our infrastructure. There was no self-service upload model like we take for granted today.

Later, by the early 2000s, however, the pieces began falling into place: faster internet connections, better video codecs, and an ever-growing hunger for on-demand media on emerging devices.

Enter YouTube.

Founded in 2005 by Steve Chen, Chad Hurley, and Jawed Karim, YouTube was the natural evolution of what we'd started with Broadcast. com. Where we aggregated and distributed live and pre-recorded professional content, YouTube democratized streaming by allowing anyone with a webcam or (later) smartphone to upload and broadcast their own videos instantly. It was a revolution that would reshape not just the internet but global culture.

How YouTube Picked Up Where Broadcast.com Left Off

In February 2005, the world's first YouTube video, a 19-second clip called *Me at the Zoo,* was uploaded, showcasing co-founder Jawed Karim talking about elephants. It was casual, unscripted, and utterly unremarkable…and it changed everything. YouTube offered a platform that was not just about consuming content but about creating it.

The timing was perfect. Broadband internet was becoming more common, video compression technology like H.264 was reducing file sizes while maintaining quality, and consumer-grade cameras were now affordable. Unlike Broadcast.com, which relied on partnerships with professional broadcasters, YouTube allowed anyone, from aspiring filmmakers to casual hobbyists, to share their voice. By then, broadband was more and more common, and bandwidth was getting cheaper.

When Google acquired YouTube in November 2006 for $1.65 billion, the platform's trajectory was cemented. From then on, YouTube wasn't just a video-sharing site; it was the foundation for a new media ecosystem.

The Rise of User-Generated Content and Its Impact on Streaming

Broadcast.com was about connecting people to live events, sports, and professional broadcasts. YouTube took it further: it connected people to each other. The rise of user-generated content, that is, videos made by everyday individuals, was marked by a seismic shift in the media landscape. Suddenly, anyone with a camera and an internet connection could reach a global audience.

The impact was profound:

- **Cultural Shifts**: YouTube birthed an entirely new class of creators. PewDiePie, one of YouTube's earliest breakout stars, started as a gaming commentator in 2010 and became a global phenomenon.
- **New Business Models**: The introduction of YouTube's Partner Program in 2007 allowed creators to monetize their videos through ad revenue, setting the stage for the influencer economy.
- **Diverse Content**: From DIY tutorials to reaction videos, YouTube's open platform encouraged niche communities and interests to flourish.

Where Broadcast.com pioneered the pre-roll ad for audio streaming, YouTube perfected it for video. Today, it's impossible to imagine a world without those five seconds you're forced to watch before you can skip to your content.

As of 2025, YouTube stands as one of the world's most-used apps, with people turning to the platform to watch live TV, tune into podcasts, and listen to music. https://www.marketwatch.com/story/youtube-now-dominates-tv-streaming-and-even-podcasts-heres-a-look-at-how-it-got-there-c8bd3bc1

Expansion into Traditional Media and Streaming

Television companies are increasingly adopting YouTube as a platform to distribute content, recognizing its prominence as a viewing medium in the U.S. Traditional media companies are leveraging their existing long-form entertainment libraries, including full-length episodes, movies, and original shows tailored for YouTube.

For instance, ITV and Fremantle have significantly expanded their content offerings on YouTube, achieving billions of views. Companies like Warner Bros. Discovery and Paramount Global's Nickelodeon are now distributing popular shows and creating original content for YouTube. With 73% of total YouTube viewing in the U.S. attributed to videos longer than 30 minutes, the platform is no longer seen solely as a hub for short user-generated clips. Media companies believe YouTube offers new advertising revenue and opportunities to engage viewers who may have missed shows on traditional TV. https://www.businessinsider.com/hollywood-tv-movies-feeding-youtube-growth-long-form-viewing-2025-2

Dominance in Podcasting

YouTube has become the most frequently used service for listening to podcasts in the U.S., surpassing platforms like Spotify and Apple Podcasts. According to data from Edison Research, 31% of weekly podcast listeners aged 13 and up reported using YouTube as their primary platform for podcasts, ahead of Spotify (27%) and Apple Podcasts (15%). This shift highlights YouTube's versatility and its ability to integrate various forms of media consumption into a single platform. https://www.marketwatch.com/story/youtube-now-dominates-tv-streaming-and-even-podcasts-heres-a-look-at-how-it-got-there-c8bd3bc1

Financial Growth and Monetization

YouTube's advertising revenue has seen substantial growth, with an estimated $36.1 billion in 2024, up from $31.5 billion the previous year. Projections suggest this figure could reach $64.2 billion by 2030. Additionally, more than 100 million subscribers have opted for services like YouTube Premium and YouTube Music, contributing significantly to the platform's revenue. This financial success underscores YouTube's effective monetization strategies and its appeal to both advertisers and consumers. https://www.marketwatch.com/story/youtube-now-dominates-tv-streaming-and-even-podcasts-heres-a-look-at-how-it-got-there-c8bd3bc1

Impact on Content Creation and Distribution

The platform's growth has also influenced content creation trends. A study analyzing YouTube Shorts versus regular videos found that creators are increasingly producing short-form content, with Shorts attracting more views and likes per view than regular videos. This trend indicates a shift in user engagement and content consumption patterns, with short-form videos becoming a significant aspect of YouTube's ecosystem. https://arxiv.org/abs/2403.00454

YouTube's evolution over the past two decades has transformed it into a central hub for diverse media consumption, impacting traditional television, streaming services, and podcasting. Its financial growth, expansive user base, and influence on content creation underscore its significant role in the modern media landscape.

The multifaceted picture of YouTube's content landscape

While YouTube boasts over 5 billion videos as of 2025, with more than 360 hours of content uploaded every minute, a significant portion of this vast library remains largely unwatched. This phenomenon, often referred to as YouTube's "ghost town," highlights that despite the platform's immense popularity, with a global user base projected to reach 2.85 billion by 2025, many videos fail to attract an audience. https://seo.ai/blog/how-many-videos-are-on-youtube, https://www.globalmediainsight.com/blog/youtube-users-statistics/

This disparity can be attributed to several factors:

- **Content Saturation**: The sheer volume of uploads makes it challenging for individual videos to gain visibility.
- **Algorithmic Influence**: YouTube's recommendation system tends to favor content that already has higher engagement, creating a cycle where popular videos become more prominent while others remain obscure.

- **Quality Variations**: User-generated content varies widely in quality and relevance, affecting viewership numbers.

Despite these challenges, YouTube continues to dominate as a leading platform for both user-generated and professional content, with over 1 billion hours of content viewed daily on TV screens alone. This duality underscores YouTube's role as a versatile medium, catering to diverse audiences and content creators, even as it grapples with the realities of content oversupply and uneven viewership distribution.

https://www.theverge.com/news/609684/youtube-bigger-tvs-phones-streaming

Innovation Beyond Broadcasting: The Deer Channel Experiment

While YouTube was redefining how people shared and consumed video, other projects were pushing the boundaries of what streaming could be. It was no longer just about broadcasting existing content or enabling user-generated videos. Streaming was becoming a tool to capture and share real-world experiences in ways that had never been possible before.

One of the more unconventional experiments in this space was Deer Channel, a project that aimed to bring audiences closer to nature by live-streaming video from the perspective of wild animals. It was a technical and logistical challenge, but it demonstrated how streaming could offer entirely new ways to experience the world.

In 2006, while streaming video was still finding its footing, I had the opportunity to work on something truly unprecedented: broadcasting life through the eyes and ears of wild deer in real-time over the internet. This wasn't just pushing technical boundaries. It was reimagining how we could experience nature itself.

The visionary project, DeerChannel.com, was the brainchild of Tom Brooks, a rancher and real estate broker who had patented a system for mounting cameras on wild animals for live internet broadcasting. When Tom approached me, the project was stuck: the contracted equipment company had delivered systems that simply didn't work.

We needed to completely re-engineer everything from the camera units to the supporting infrastructure, including a fiber optic ring around the habitat, PTZ cameras on towers, and receivers in weather-hardened NEMA enclosures.

The technical challenges were immense, especially considering this was during the pre-iPhone, pre-GoPro era. I cobbled together solutions using off-the-shelf spy equipment (yes, literally from a place called the Spy Store) and custom-engineered protective housing. With an unlikely combination of help from a saddle maker, a blacksmith, and a plumber, through successive iterations, we created a camera system so robust you could drive a truck over it, which we actually did, just to prove the point.

But the real complexity came from our subjects. We were working with wild deer, which meant dealing with strict biological constraints. Male deer only keep their antlers for a few months each year, giving us a narrow window for deployment. The process was intricate: dart the deer (which I learned required weeks between attempts due to health concerns), mount the equipment carefully on their antlers (after they harden) to avoid harm, pump the animal full of antibiotics (darting apparently compromises their immune system), and release them back into the wild.

Here's the catch: you can't exactly schedule maintenance or recharge once they're free. The deer's first instinct was to try knocking the equipment off by ramming their antlers against trees, a natural behavior that destroyed our first several versions. The solution? We encased the spy cameras in Schedule 80 PVC pipe with a bulletproof glass front.

For power management, we used the best lithium-ion batteries available at the time, but even those had limits. Once the battery was dead, that was it. No more signal.

The original plan had been to use a "solar blanket" that the deer would have draped over its back to charge the batteries. Unfortunately, having something on the deer's back caused an instinctive reaction as though it were being attacked.

We implemented a timer system to activate only during peak deer activity hours, extending our operational window significantly. I consulted with my friend, the late Jim Cline, an engineer, to build the timer. He ran into difficulties researching this because, well, after 9/11 and concerns over terrorism, there were certain sensitivities to the engineering of timers that raised alarms as he talked to people online trying to find existing designs.

There were off-the-shelf systems like you would find in ordinary things like water timers, but things we found were too large or in various ways wouldn't be suitable for what we needed – something very small, lightweight, and could be inserted into a tube… oh… well… now that I'm remembering, I can see where Jim ran into issues. Long story short, he got something working for us.

The results from DeerChannel were extraordinary. On our very first day of operation, we captured footage that contradicted established wildlife biology textbooks – our deer waded into a pond and started eating aquatic vegetation, behavior that wasn't supposed to exist according to the literature. We captured incredible moments, including infrared footage of two bucks engaged in antler-to-antler combat at night. The audio was just as revealing as the video, offering an unprecedented window into how deer perceive their world.

At the time of this writing, there was still a low-rez Deer Channel promo clip on YouTube (see link below). And, no, there was no hunting associated with Deer Channel itself. No deer were targeted, etc., despite some of the content in the clip. https://www.youtube.com/watch?v=qZrPeg4g124

The channel wasn't just about cool footage, though. The project was overseen by university biologists, focusing on whitetail deer genetics and habitat optimization. We conducted controlled experiments, like planting different crops side-by-side to study deer feeding preferences, all aimed at breeding superior Boone and Crockett bucks.

We monetized through pay-per-view streaming, merch, and documentary production and were in discussions with major nature channels about broader distribution. We even had plans to take the concept to Africa.

While Deer Channel didn't ultimately survive due to circumstances beyond our control, it demonstrated something important about streaming's potential: with enough innovation and determination, we could create entirely new ways of experiencing the natural world. Today's wildlife documentaries and live nature feeds owe a little something to those early experiments with deer-mounted cameras in 2006.

The Shift from Centralized Platforms to Decentralized, User-Driven Media

YouTube may have been the first major player to decentralize content creation, but it certainly wasn't the last. In the years following its launch, platforms like **Twitch** (founded in 2011) and **TikTok** (launched in 2016) continued to push the boundaries of user-driven media.

Twitch

Originally a gaming-focused offshoot of Justin.tv, revolutionized live-streaming by creating a space where real-time interaction between creators and viewers was as important as the content itself. It started as a hub for gaming enthusiasts, but over time, it expanded into a broader entertainment platform featuring everything from esports tournaments to live music performances and even "just chatting" streams where creators engage directly with their audiences.

As of January 2025, Twitch attracts over 240 million unique visitors per month, with an average of 2.5 million concurrent viewers at any given time. The platform continues to dominate the live-streaming space, though it has faced competition from platforms like YouTube Live and Kick.

https://backlinko.com/twitch-users

While Twitch saw its peak number of active streamers at 9.89 million in January 2021, that number has since settled to 7.7 million active streamers per month in 2025. The platform remains a leader in gaming content, but its evolution into lifestyle, talk shows, and creative arts has cemented its place as a major player in the modern streaming landscape. https://www.statista.com/statistics/746173/monthly-active-streamers-on-twitch/

In the U.S. alone, Twitch is projected to have 35.6 million users in 2024, accounting for over 20% of its global audience. This sustained growth highlights Twitch's ability to adapt, keeping audiences engaged while continuing to shape the future of live digital entertainment. https://www.demandsage.com/twitch-users/

TikTok

Launched globally in 2018 following the merger of Musical.ly and Douyin (its Chinese counterpart), TikTok quickly became a dominant force in the social media and streaming landscape. By emphasizing short-form, mobile-first video content, TikTok captured the imagination of Gen Z and beyond, transforming passive viewing into an interactive, algorithmically curated experience. Its "For You" page, powered by one of the most advanced recommendation engines in the industry, made content discovery frictionless, addictive, and deeply personalized, even for users who never followed a single account.

What sets TikTok apart is its democratization of reach. Creators with no prior following can go viral overnight, upending the traditional influencer hierarchy established on platforms like YouTube and Instagram. TikTok's format encourages trends, challenges, and duets that create a constantly evolving shared culture among users. The comment sections often become their own ecosystems of humor, debate, and interaction, blurring the line between audience and creator.

In the live-streaming space, TikTok Live has grown rapidly as a competitor to Twitch and Instagram Live, especially in music, shopping, and behind-the-scenes content. TikTok has also made aggressive moves into e-commerce, integrating shopping features directly into the platform, positioning itself as not just a media app, but a full-fledged entertainment-commerce hybrid.

As of early 2025, TikTok boasts more than 1.6 billion active users globally, with over 150 million users in the U.S. alone. Its cultural impact is so pervasive that music charts, product trends, and even television programming are often shaped by what gains traction on the app. While it differs from Twitch in format and function, TikTok's influence on the modern streaming economy, especially in mobile, creator monetization, and real-time engagement, is undeniable.

Lessons from Netflix, Twitch, and the Democratization of Streaming

While YouTube democratized content creation, Netflix demonstrated the power of streaming at scale. Founded in 1997 as a DVD rental service, Netflix mailed DVDs directly to customers' homes, allowing them to watch and return them at their convenience. When Netflix transitioned from DVD rentals to streaming in 2007, it was a gamble that paid off in spades.

By 2013, with the launch of House of Cards (don't blame me if the theme song is playing in your head now…), Netflix had firmly established itself as a leader in original programming. Its strategy of licensing, producing, and distributing content directly to consumers bypassed traditional networks entirely. Though Netflix officially stopped offering DVD rentals in 2023, the legacy of its transition to streaming remains a pivotal moment in media history.

Meanwhile, Twitch proved that niche platforms could thrive by serving specific communities. Its focus on gaming gave rise to the phenomenon of live-streaming, where interaction between creators and viewers became as important as the content itself. Together, these platforms highlighted a critical lesson: streaming isn't just about delivering content; it's about building communities.

This shift in the media landscape has empowered independent creators and "citizen reviewers" to rival traditional industry voices in influence and reach. Platforms like YouTube and Twitch have amplified voices outside the mainstream, giving rise to creators (think: MrBeast (364 million), PewDiePie (111 million), Markiplier (36.8 million), Joe Rogan (19.4 million),) whose (often stridently) unfiltered opinions and direct engagement resonate with audiences.

Unlike traditional outlets such as Variety or The Hollywood Reporter, which are often perceived as beholden to the entertainment industry, these independent voices offer candid critiques that audiences increasingly trust. This transparency has not only shaken the influence of traditional media but has also forced it to adapt, incorporating elements of the direct, personal style that defines independent content.

Nowhere is this more evident than in the realm of pop culture and gaming reviews. Popular YouTubers and Twitch streamers routinely amass millions of views on their reactions, reviews, and breakdowns of movies, games, and shows. Their reach often eclipses that of traditional media outlets, thanks to their ability to cultivate deeply loyal, engaged communities.

In many ways, they have redefined what it means to be a critic, blending entertainment, expertise, and direct audience interaction into a powerful and increasingly influential format. This democratization of media criticism exemplifies how streaming platforms have not only transformed how we consume content but also how we interpret and discuss it.

However, this democratization hasn't been without challenges. Issues like so-called "misinformation," outright censorship, and copyright disputes have plagued YouTube and its successors. The "Adpocalypse" of 2017, when advertisers pulled out due to controversial content, was a stark reminder that the balance between freedom and responsibility in user-driven platforms remains fragile.

Legacy of Broadcast.com in the Post-YouTube Era

When I look at platforms like YouTube, Netflix, and Twitch, I see the echoes of Broadcast.com everywhere. Our work to solve early challenges in live streaming, from bandwidth optimization to content delivery networks, laid the groundwork for what these giants have achieved.

Broadcast.com showed that the internet could be more than text and static images – that it could connect people in real-time, whether to a football game or a business conference. But while Broadcast.com laid the foundation, Yahoo!, which acquired us in 1999, had a five-year head start to dominate the streaming space and blew it. Instead of capitalizing on the infrastructure and vision we'd built, Yahoo! struggled to integrate our technology and squandered the opportunity to lead the industry. It's a cautionary tale of how strategic missteps and lack of focus can derail even the most promising beginnings.

YouTube took the idea of connecting people through media and turned it into a global phenomenon, proving that anyone, anywhere, could share their story. By embracing user-generated content and fostering a community-centric platform, YouTube succeeded where Yahoo! faltered, showing the world what streaming could truly become.

The Future of Streaming

Streaming has come a long way since the days of dial-up modems and RealAudio codecs. When we launched Broadcast.com, the goal was to take traditional media, radio, and, later, television, and live events and bring them to the internet, making them accessible to anyone with a connection. We were building a new way to distribute professional broadcast content long before the technology had fully caught up.

YouTube took a different path. Instead of focusing on existing broadcasters, it gave the tools of distribution to everyday users. It transformed the internet into a global stage where anyone could create and share content, shifting the power from networks and studios to individuals. This shift reshaped not just media consumption but also culture, launching the creator economy and making influencers as relevant as traditional celebrities.

Now, the cycle has come full circle. The same mainstream networks and studios that once controlled distribution are embracing YouTube as their primary platform. Television companies, movie studios, and even sports leagues are leveraging YouTube's reach, using it not just for clips and marketing but for full-length programming, live broadcasts, and exclusive content. What started as a platform for user-generated videos is now one of the largest distribution channels for traditional media.

The evolution of streaming isn't just about technology. It's about who controls distribution, and that answer keeps changing. We started with legacy broadcasters, then shifted to independent creators, and now, media giants are reclaiming space on digital platforms. The lines between traditional and new media are blurring, and the next era of streaming will be defined by how these worlds continue to merge.

Through it all, the core of streaming has remained the same: bringing content to audiences wherever they are. That was the mission of Broadcast.com. It was the promise of YouTube. And it will shape whatever comes next.

What are your thoughts on how far streaming has come? Did you watch the first YouTube videos, or were you tuning into live events on Broadcast.com?

9: From Broadcast.com to Global Streaming

The Protocols and Technologies Powering Today's Platforms

Streaming has transformed how we consume content, evolving from clunky RealAudio streams on dial-up connections to today's seamless 4K and live interactive experiences. While consumer-facing innovations grab the headlines, the real revolution has been in the underlying technologies: compression codecs, networking protocols, and transport mechanisms. These advancements have quietly reshaped the internet, enabling the scale, speed, and interactivity we now expect.

This chapter explores the technologies that power streaming platforms like YouTube, Netflix, Twitch, and more, highlighting how each leverages protocols and transports to meet specific needs.

Streaming platforms employ a variety of protocols and transport methods, carefully chosen to address specific priorities such as latency, reliability, and scalability. These technical decisions play a critical role in shaping the viewer experience across different types of content and use cases.

YouTube, for instance, leverages HTTP Live Streaming (HLS) and Dynamic Adaptive Streaming over HTTP (DASH) protocols alongside the Transmission Control Protocol (TCP). These technologies enable smooth, adaptive playback for on-demand and live content by adjusting video quality to match the viewer's network conditions. TCP ensures reliable delivery of every data packet, making it ideal for a platform that prioritizes consistency over ultra-low latency.

Similarly, Netflix employs DASH combined with proprietary optimizations, also using TCP as the transport layer. Netflix's focus on high-quality video and seamless playback across various network conditions underscores its commitment to ensuring a premium viewing experience. Adaptive bitrate streaming with DASH allows Netflix to dynamically scale video quality, even in less-than-optimal conditions, preserving the integrity of its content.

Twitch, a live streaming platform known for its real-time interactivity, employs different strategies to balance latency and reliability. For content ingest, Twitch relies on Real-Time Messaging Protocol (RTMP), which uses the User Datagram Protocol (UDP) for low-latency uploads from streamers. For end-user playback, it employs HLS over TCP, which ensures stream stability and compatibility across devices.

This dual approach enables Twitch to cater to both the real-time needs of streamers and the stable viewing experience expected by its audience. In a similar vein, Zoom prioritizes real-time communication in video conferencing by using Real-Time Transport Protocol (RTP) over UDP. This combination prioritizes speed over error correction, allowing conversations and video feeds to remain fluid, even if some packets are lost along the way.

On the other hand, Spotify, a music streaming platform, relies on proprietary protocols built on TCP. Music streaming demands high fidelity and reliability, making TCP an excellent choice as it ensures every packet of audio data is delivered accurately, preserving the quality of the listening experience.

Disney+ and Amazon Prime Video, much like Netflix, focus on delivering high-quality, long-form content. These platforms also use HLS and DASH protocols with TCP as the transport layer. By prioritizing reliability and adaptive bitrate streaming, they can offer uninterrupted playback, even in fluctuating network conditions.

Each platform's protocol and transport choices reflect its core mission and audience needs. Whether prioritizing real-time interaction, as seen with Twitch and Zoom, or ensuring consistent quality for on-demand content like Netflix and Disney+, these technical decisions are integral to the success of modern streaming services. As consumer expectations continue to evolve, these choices will remain pivotal in shaping the future of digital content delivery.

Summary:

1. YouTube (On-Demand and Live Streaming):
   - Protocols Used: HLS (HTTP Live Streaming) and DASH (Dynamic Adaptive Streaming over HTTP)
   - Transport: TCP (Transmission Control Protocol)
   - Why: YouTube primarily delivers on-demand content, where buffering is less critical than smooth playback. TCP ensures reliable delivery, and HLS/DASH enables adaptive bitrate streaming, adjusting quality based on network conditions.
2. Netflix (On-Demand Streaming):
   - Protocols Used: DASH and proprietary implementations optimized for efficiency.
   - Transport: TCP.
   - Why: As an on-demand platform, Netflix focuses on high-quality video. TCP guarantees packet delivery, ensuring no loss in quality. Adaptive bitrate streaming with DASH allows seamless playback, even in variable network conditions.
3. Twitch (Live Streaming):
   - Protocols Used: RTMP (Real-Time Messaging Protocol) for ingest, HLS for playback.
   - Transport: UDP for low-latency streams during live ingest; TCP for stable end-user playback.
   - Why: Twitch prioritizes real-time interactivity, making low-latency ingest critical. RTMP over UDP enables fast, stable uploads, while HLS over TCP ensures reliability for viewers, albeit with a slight delay.
4. Zoom (Video Conferencing):
   - Protocols Used: RTP (Real-Time Transport Protocol)
   - Transport: UDP
   - Why: Real-time communication demands minimal latency, making UDP and RTP ideal. These protocols prioritize speed over reliability, as packet loss in live communication is less disruptive than delays.
5. Spotify (Music Streaming):
   - Protocols Used: Proprietary streaming protocols built

on TCP
- o Why: Music streaming requires high fidelity and reliability. TCP ensures every packet is delivered, preserving audio quality even in poor network conditions.

6. Disney+ and Amazon Prime Video (On-Demand):
   - o Protocols Used: HLS and DASH
   - o Transport: TCP
   - o Why: Like Netflix, these platforms prioritize quality and reliability for long-form content. TCP is well-suited for their focus on consistent playback without errors.

Hardware

Back in the late 1990s, when we were buying Dell OptiPlex computers as encoders and server machines for streaming, we were running Windows NT on them. These machines were not cheap. Each one cost us somewhere between $2,000 and $3,000, and that was before adding a Sound Blaster card or any additional upgrades. These were enterprise-class desktops that had to do double duty, not just as office workstations but as the backbone of our streaming infrastructure. At the time, encoding was an expensive and CPU-intensive process, and we were pushing these machines to their limits.

Today, that same functionality, which once required a dedicated server, can be handled by a $300 laptop. For a radio station streaming setup, I recently ordered a modern notebook along with an external audio interface and supporting equipment, all for a fraction of the cost. The power and efficiency of modern hardware, even at the consumer level, would have been unimaginable back in our Broadcast.com days. Of course, in the very early days, we were stuck using el-cheapo Packard Bell or whatever PCs we could scrounge.

The real shift today, however, is not just in hardware. It is in the way compute resources are deployed.

Back then, everything was on-premise. There was no cloud.

We physically racked servers, managed cooling backup power, and sweated over bandwidth costs. Today's equivalent would likely be a cloud-based setup, leveraging scalable virtualized infrastructure, distributed encoding, and AI-assisted optimizations. That shift, from individual workstations grinding away on media encoding to cloud-distributed, GPU-accelerated computing, has fundamentally transformed how streaming operates. Our OptiPlex systems were the best we could get at the time, but by today's standards, they were relics of an era when local processing was the only option.

Even more than hardware costs, the efficiency gains have been staggering. A single GPU in a cloud data center can now do in seconds what our old systems would have taken ages to accomplish. The move from manually managing servers to serverless compute environments, content-aware encoding, and AI-assisted media processing is as big a leap as the one from dial-up to fiber. If we had access to this kind of infrastructure in the 1990s, Broadcast.com might have scaled even at a fraction of the cost. On the other hand, more and more companies are moving their servers back into on-premise or company-operated data centers.

Compression: The Heart of Streaming Efficiency

Compression codecs are fundamental to streaming platforms, enabling efficient transmission of high-quality media by balancing file size and fidelity.

The Beginnings: MPEG-2 and Early Challenges

In the late 1990s and early 2000s, MPEG-2 was the standard codec for DVDs and early digital broadcasts. Its compression capabilities were revolutionary for the time, allowing media to move from bulky physical formats to digital. However, as internet streaming began to emerge, MPEG-2's large file sizes and inefficiencies became a bottleneck. It paved the way for a new era of codecs designed for online video.

The Reign of H.264: Ushering in the Streaming Era

Released in 2003, H.264 (AVC) quickly became the cornerstone of the digital video revolution. Its ability to deliver high-quality video at significantly reduced file sizes made streaming viable for platforms like YouTube, Netflix, and Vimeo. Its wide compatibility across devices and operating systems cemented its dominance, and it remains the most widely supported codec even today.

The development of H.264 wasn't just technical. It was a response to the explosive growth of video on the internet. As YouTube launched in 2005 and Netflix pivoted to streaming in 2007, the need for a codec that balanced quality, efficiency, and compatibility became critical. By the late 2000s, H.264 enabled everything from 720p videos on YouTube to HD movie rentals on iTunes, redefining how media was consumed globally.

H.265 and the Rise of 4K: Efficiency Meets High-Quality Video

As HD became the norm and 4K content emerged, the limitations of H.264 became apparent. H.265 (HEVC) was introduced in 2013, offering up to 50% better compression efficiency than H.264. This made it possible to stream ultra-high-definition content on platforms like Netflix and Amazon Prime Video without massive bandwidth requirements.

However, H.265 faced challenges. Its licensing complexities and high royalties discouraged universal adoption. While giants like Apple and Netflix embraced it for high-end content, many smaller platforms hesitated, opening the door for newer, open-source competitors.

The AV1 Revolution: A Royalty-Free Future

The next chapter in the codec story emerged with AV1, developed by the Alliance for Open Media (AOMedia), a consortium of industry heavyweights including Google, Netflix, Amazon, and Microsoft. Launched in 2018, AV1 promised superior compression efficiency (20-30% better than H.265) without the licensing fees. This open-source approach was a game-changer, allowing platforms to innovate without incurring prohibitive costs.

YouTube and Netflix were early adopters of AV1, streaming select content to users with compatible devices. The codec's efficiency and cost-effectiveness have positioned it as a strong contender for the future of streaming, especially as 8K video and immersive technologies like AR and VR become mainstream.

The Evolution of Audio Codecs: From MP3 to Opus and FLAC

While video codecs revolutionized visual storytelling, audio codecs have undergone a parallel transformation, shaping the way we consume music, podcasts, and real-time voice communication. The journey from MP3 to AAC, Opus, and FLAC reflects the trade-offs between compression, quality, and efficiency, each optimized for different use cases.

MP3: The Codec That Changed Music Forever

In the 1990s and early 2000s, MP3 (MPEG-1 Audio Layer III) became the dominant format for digital audio. Its lossy compression dramatically reduced file sizes, making it possible to store thousands of songs on early iPods and stream music over dial-up connections. Despite some quality loss, MP3 was revolutionary, enabling the rise of platforms like Napster, iTunes, and early streaming services. However, as internet speeds improved and audio quality expectations grew, newer formats emerged to offer better efficiency and fidelity.

AAC: The Successor to MP3

AAC (Advanced Audio Coding), introduced in the late 1990s, offered better sound quality at lower bitrates compared to MP3. It became the default codec for iTunes, YouTube, and Apple Music, as well as many streaming video platforms. Unlike MP3, AAC provided improved efficiency and multi-channel support, making it a superior choice for modern media applications.

FLAC: The Gold Standard for Lossless Audio

For audiophiles and high-fidelity audio applications, FLAC (Free Lossless Audio Codec) emerged as the preferred format. Unlike MP3 and AAC, which discard some audio data during compression, FLAC maintains 100% of the original audio quality while reducing file size by 30-50%. This makes it ideal for archiving music, professional audio production, and lossless streaming. Platforms like Tidal, Qobuz, and Bandcamp support FLAC for users who prioritize sound quality over storage constraints. However, due to its larger file sizes, FLAC is less suited for bandwidth-limited streaming and real-time applications.

Opus: Optimized for Real-Time Streaming & Communication

While MP3, AAC, and FLAC cater to music and media playback, Opus was designed for low-latency, high-quality voice and audio streaming. Developed in 2012, Opus can dynamically adjust bitrate, latency, and audio quality based on network conditions, making it perfect for voice-over-IP (VoIP), gaming, and video conferencing. It is widely used in platforms like Discord, Zoom, WhatsApp, and WebRTC-powered applications. Unlike MP3 or AAC, Opus is optimized for both speech and full-bandwidth music, making it a versatile choice for real-time communication.

Choosing the Right Audio Codec

Each of these codecs serves a different purpose:
- MP3: Still widely used for its compatibility and small file sizes, but now considered outdated for high-quality streaming.
- AAC: The modern standard for music streaming and video playback offers better compression and quality than MP3.
- FLAC: The best choice for lossless, high-fidelity audio used by audiophiles and professional musicians.

- Opus: The go-to codec for real-time applications, ensuring low latency and adaptive quality audio for VoIP, gaming, and conferencing.

As streaming technology evolves, these audio codecs will continue to shape how we listen, communicate, and experience sound in the digital age. Whether you are enjoying a lossless FLAC album, streaming a song in AAC, or chatting on Discord using Opus, the underlying codec plays a crucial role in delivering the best possible audio experience.

A Future of Innovation and Efficiency

The evolution of codecs continues to mirror the changing landscape of digital media. With the rise of 8K video, live-streaming platforms, and virtual reality, codecs like VVC (Versatile Video Coding) and EVC (Essential Video Coding) are on the horizon, promising even greater efficiency. The codec journey is far from over, as industry players balance the demands of cutting-edge quality, accessibility, and cost.

Network Protocols: Choosing Speed vs. Reliability

Streaming technologies are shaped by the delicate balance between quality, speed, and scalability. Each use case, whether it's binge-watching a Netflix series, live-tweeting during a Twitch stream, or participating in a Zoom call, requires a unique approach to delivering content. The evolution of these methods tells a story of relentless innovation, where engineers have consistently pushed the limits to match viewer expectations and technological capabilities.

The Evolution of Trade-offs in Streaming Protocols

The Early Days: Quality Over Speed

In the early days of the internet, when streaming was in its infancy, platforms prioritized reliability. Dial-up connections couldn't handle large media files in real-time, and buffering was a familiar annoyance. TCP (Transmission Control Protocol) emerged as the backbone of early streaming due to its ability to ensure every data packet arrived intact. Packets are reassembled at the destination, sometimes having arrived via disparate paths. As a result, latency can be significant; however, the method helps ensure the quality of the resulting playback (even if it might be as long as minutes delayed). This reliability made it indispensable for platforms like YouTube and Spotify as they set the stage for on-demand content.

Real-Time Streaming

The need for real-time, no-lag streaming has become critical for certain vertical applications where even slight delays can break the experience. In-venue live sports broadcasting, financial trading platforms, remote surgery, and online multiplayer gaming all require near-instantaneous data transmission. In esports, for example, platforms like Twitch and YouTube Gaming must minimize latency to maintain engagement between streamers and audiences. Similarly, financial markets rely on ultra-low latency data feeds for high-frequency trading, where milliseconds can mean millions of dollars. In telemedicine, real-time video and haptic feedback are crucial for remote robotic surgeries, ensuring that surgical commands translate instantly without lag.

To meet these demands, streaming protocols evolved to prioritize speed over absolute reliability. UDP (User Datagram Protocol) and RTP (Real-Time Transport Protocol) became essential, allowing for ultra-fast data transmission by minimizing delays, even if a few packets were lost along the way. Unlike TCP, which ensures every packet arrives before processing, these protocols enable real-time interactions by transmitting data continuously, making them the backbone of industries where responsiveness is non-negotiable.

Modern Scalability and Adaptability

Today, protocols like HLS (HTTP Live Streaming) and DASH (Dynamic Adaptive Streaming over HTTP) combine the strengths of earlier technologies, ensuring high quality while adapting to fluctuating network conditions. The rise of multicasting also offers promising solutions for enterprise and large-scale live events, optimizing bandwidth usage for massive audiences.

Summary of Key Protocols and Their Trade-offs

1. TCP (Reliable and Predictable): TCP ensures that every data packet sent is delivered and reassembled in the correct order. This makes it the gold standard for on-demand content where quality and reliability are paramount. Platforms like Netflix, YouTube, and Spotify rely on TCP for smooth, uninterrupted playback. Its trade-off? Latency. The process of checking and re-sending missing packets introduces delays, which is acceptable for on-demand streaming but unsuitable for real-time applications.

2. UDP (Fast but Unreliable): UDP trades reliability for speed, making it ideal for live streaming and interactive platforms where delays can disrupt the experience. Unlike TCP, UDP doesn't wait for an acknowledgment that packets have been received. This allows for lower latency but comes at the cost of occasional packet loss. Twitch, Zoom, and live sports platforms embrace UDP to ensure real-time delivery, even if a frame or two is dropped.

3. RTP (Layered Over UDP): RTP builds on UDP by adding features to manage timing and ordering, essential for live communication and streaming. This protocol enables real-time delivery with minimal latency, even if some packets are lost. Applications like Zoom, Microsoft Teams, and WebRTC (used in Google Meet) depend on RTP to provide seamless audio and video for calls and meetings.

4. Multicasting: Multicasting sends data to multiple recipients simultaneously, reducing the bandwidth required for live broadcasts. While underutilized in consumer platforms, it is crucial for enterprise IPTV systems and large-scale multicast sports streams. Multicasting is highly efficient but requires robust infrastructure and isn't widely supported by home networks or devices, limiting its application outside specialized use cases.

5. HLS and DASH: HLS and DASH break video streams into

small segments, allowing platforms to adjust the quality of playback dynamically based on the viewer's network conditions. These protocols are widely adopted by platforms like YouTube, Disney+, and Amazon Prime Video for their scalability and compatibility. They excel at providing high-quality playback but often introduce a latency of 10-30 seconds, making them less ideal for live or real-time applications.

The Future of Streaming Protocols

As streaming continues to evolve, hybrid approaches are becoming the norm. Platforms increasingly mix technologies, combining the reliability of TCP with the speed of UDP or leveraging edge computing to reduce latency in HLS streams. Emerging use cases like AR, VR, and real-time sports betting are driving the need for innovations that push the boundaries of these protocols.

The story of streaming isn't just about technical trade-offs. It's about meeting the expectations of a global audience. Whether it's ensuring that the final scene of a movie arrives in perfect quality or delivering a live goal with zero delay, the ongoing evolution of streaming technologies ensures that every moment, however fleeting, is delivered with precision.

The Evolution of CDNs: From Broadcast.com to Modern Edge Networks

When we were building Broadcast.com in the 1990s, one of our biggest challenges was delivering content to a growing, geographically dispersed audience with the limited bandwidth (mostly dialup) and infrastructure of the time. To solve this, we developed what I believe was one of the first Content Delivery Networks (CDNs), even if we didn't call it that back then.

At the time, our servers and systems were centralized, creating latency and bottlenecks for users far from our Dallas data center. Buffering was rampant, and bandwidth costs were exorbitant, especially for major events like the MLB World Series or live corporate earnings calls. We had to innovate to meet demand while staying cost-effective.

Broadcast.com: The Early CDN Blueprint

Our solution was to create distributed nodes, essentially edge servers, positioned in strategic locations across the U.S. and eventually globally. We set up servers closer to end-users, reducing the physical distance data had to travel and cutting latency. Satellite uplinks and dedicated telco lines fed live content to these nodes, which then streamed it to local audiences.

Of course, for a while, we even had different links our audience could use to find servers that were fewer "hops" away from them.

To handle live broadcasts, we also installed an array of satellite dishes at our facility, making Broadcast.com one of the largest satellite farms in the Southwest. These dishes ingested feeds from various broadcasters and distributed them to our server network. It wasn't perfect; we still had limitations in bandwidth and redundancy, but it was a revolutionary step toward scalable, real-time content delivery.

Our approach foreshadowed the modern CDN by prioritizing the following:

1. **Latency Reduction:** Minimizing delays by bringing content closer to users.
2. **Scalability:** Allowing us to handle spikes in demand for high-profile events.
3. **Redundancy:** Ensuring streams remained active even during hardware or network failures.

The Birth of Modern CDNs

As internet infrastructure matured in the early 2000s, companies like Akamai formalized the CDN concept. Their platforms expanded on the principles we pioneered, using intelligent routing, caching, and redundancy to deliver content efficiently.

Key milestones in the evolution of CDNs:

- **2000s – Caching and Global Reach:** Akamai introduced caching at edge servers, storing frequently accessed content locally to reduce latency. Their global network laid the foundation for widespread CDN adoption.

- **2010s – Adaptive Bitrate Streaming:** Platforms like Netflix and YouTube leveraged CDNs to deliver adaptive bitrate streams. These systems automatically adjust video quality based on a user's connection, ensuring a seamless experience regardless of bandwidth variability.
- **2020s – Edge Computing and AI:** Modern CDNs incorporate edge computing, processing data closer to users for ultra-low latency applications like AR/VR and online gaming. AI optimizes content delivery, predicting demand and pre-caching content.

CDNs in Action: Powering Today's Streaming Platforms

Today, every major streaming platform relies on CDNs to deliver content at scale:

- **Netflix:** Operates its own Open Connect CDN, delivering terabytes of data daily from strategically placed edge servers.
- **YouTube:** Uses Google's global CDN infrastructure to manage billions of daily views.
- **Twitch:** Relies on Amazon CloudFront for real-time, low-latency delivery of live streams.
- **Disney+:** Combines public CDNs with private caching to optimize video delivery for global audiences.

Beyond video, CDNs power everything from Spotify's music streaming to e-commerce platforms like Amazon and even multiplayer online games.

Why CDNs Matter More Than Ever

In today's digital landscape, Content Delivery Networks (CDNs) are more critical than ever, especially as streaming becomes increasingly global and real-time demands intensify. Latency-sensitive applications such as live sports, online gaming, and immersive AR/VR experiences rely heavily on advanced CDN features to ensure optimal performance.

Low-Latency Delivery: For live events and real-time interactions, minimizing latency is paramount. CDNs achieve this by distributing content across globally dispersed servers and caching data at edge locations near end users. This proximity reduces the time data travels, resulting in faster load times and a more responsive experience. For instance, Gcore's next-generation CDN delivers a global response time of less than 30 milliseconds, enabling dynamic content stream acceleration. https://www.sportspro.com/insights/gcore-cdn-tech-streaming-ultra-low-latency/

Geographically Distributed Edge Networks: To maintain high performance, even in remote regions, CDNs deploy servers worldwide, bringing content closer to users. Akamai's Intelligent Edge Platform exemplifies this approach, operating approximately 365,000 servers in over 135 countries. This extensive network gathers real-time information about traffic and congestion, ensuring efficient content delivery regardless of user location. https://en.wikipedia.org/wiki/ Akamai_Technologies

Dynamic Caching: CDNs employ dynamic caching strategies to adapt to user behavior and network conditions in real-time. By temporarily storing copies of files in data centers across the globe, CDNs can deliver content from the server closest to the user, reducing page load times and enhancing the web experience. https://www.akamai. com/glossary/what-is-a-cdn

In summary, as the demand for instantaneous, high-quality content grows, CDNs continue to evolve, incorporating advanced features to meet the needs of latency-sensitive applications. Their role in delivering seamless experiences across various platforms underscores their indispensable place in the modern internet ecosystem.

- **Low-Latency Delivery:** Critical for live events and real-time interactions.
- **Geographically Distributed Edge Networks:** Ensuring high performance even in remote regions.
- **Dynamic Caching:** Adapting to user behavior and network conditions in real-time.

Reflections on the Road Ahead

Looking back, the early experiments at Broadcast.com were more than just stopgap measures. They were the beginning of a paradigm shift in content delivery. We laid the groundwork for what would become a multi-billion-dollar industry, enabling today's seamless streaming experiences.

As technologies like edge computing and AI redefine what's possible, the principles we established, scalability, latency reduction, and redundancy, will continue to drive innovation. CDNs aren't just the hidden backbone of streaming; they're the unsung heroes of the digital age.

Latency: The Industry's Persistent Challenge

Latency, the delay between a live event and its broadcast to viewers, remains a significant challenge in live streaming. Despite advancements in compression and streaming protocols, platforms like YouTube and Twitch often experience delays of 20 to 30 seconds behind real-time events. This lag results from necessary processes such as buffering, encoding, and transmission designed to ensure smooth playback.https://support.google.com/youtube/answer/7444635

Emerging Solutions:

Low-Latency HLS (HTTP Live Streaming)

Apple's Low-Latency HLS extension aims to reduce delays to 2-3 seconds by introducing partial media segments and utilizing HTTP/2 for efficient data transmission. However, implementing Low-Latency HLS requires optimized network infrastructure and support across various devices and platforms. https://en.wikipedia.org/wiki/HTTP_Live_Streaming

WebRTC (Web Real-Time Communication)

WebRTC is a powerful technology designed for peer-to-peer communication, enabling near-instantaneous delivery that is ideal for applications like video conferencing. However, when it comes to scaling WebRTC for large audiences, challenges arise due to its inherent architecture.

In its native form, WebRTC is optimized for direct peer-to-peer connections, which works seamlessly for one-on-one or small-group interactions. As the number of participants increases, especially into the hundreds or thousands, the scalability becomes a concern. Each additional participant requires a separate connection, which can strain the broadcaster's upload bandwidth and processing capabilities. For instance, streaming to 10,000 attendees would necessitate 10,000 individual streams, which is impractical for most setups. https://stackoverflow. com/questions/18318983/webrtc-scalable-live-stream-broadcasting-multicasting

To overcome these limitations, solutions like media servers or selective forwarding units (SFUs) are employed. These servers receive a single stream from the broadcaster and distribute it to multiple viewers, effectively managing the load and enabling scalability to larger audiences. With such architectures, streaming to tens or even hundreds of thousands of viewers has been demonstrated successfully. https:// ceeblue.net/webrtc-scalability/

However, scaling WebRTC to audiences in the millions presents significant challenges and is not typically feasible with standard implementations. Alternative streaming protocols or hybrid solutions are often considered for such large-scale broadcasts to ensure reliability and performance.

CMAF (Common Media Application Format)

CMAF is transforming how platforms approach the challenges of latency and scalability in live streaming. Introduced by Microsoft and Apple, CMAF is designed to standardize media delivery, reduce latency, and maintain compatibility across a broad range of devices and platforms. This makes it particularly appealing for applications that require real-time engagement, such as live sports, betting, and interactive entertainment.

How CMAF Works

At its core, CMAF is a media file format optimized for HTTP-based adaptive streaming protocols, specifically HLS (HTTP Live Streaming) and DASH (Dynamic Adaptive Streaming over HTTP).

Its most significant contribution lies in its support for chunked transfer encoding, where video segments are divided into smaller, manageable chunks. Instead of waiting for an entire segment to be encoded and transmitted, chunks can be streamed to viewers as soon as they are ready. This approach reduces latency while maintaining the quality and consistency of playback.

CMAF also reduces redundancy by using a single media file that can be delivered through both HLS and DASH. Previously, platforms needed to create separate files for each protocol, resulting in higher storage and processing overhead. By unifying these workflows, CMAF not only improves efficiency but also enhances scalability for streaming platforms.

Applications of CMAF

Live Sports and Betting: For live sports platforms like ESPN+ or DAZN, even a few seconds of latency can significantly impact user engagement, especially when fans rely on second-screen interactions or live betting. CMAF's low-latency capabilities ensure viewers receive content almost in sync with the live action, keeping the experience immersive and interactive.

Interactive Features: Platforms incorporating features such as live polls, chat, or synchronized viewing sessions benefit from CMAF's real-time delivery. By reducing delays, these features feel more natural and engaging to users.

Immersive Technologies: Emerging formats like AR (Augmented Reality) and VR (Virtual Reality) demand ultra-low latency to provide seamless experiences. CMAF's efficient chunking and delivery mechanisms make it a suitable choice for these cutting-edge applications.

Why Latency Matters

Latency remains one of the most critical factors in live streaming. Traditional live streams on platforms like YouTube and Twitch often experience delays of 20-30 seconds due to processes like encoding, buffering, and content delivery. While this delay is tolerable for

STREAMING WARS

on-demand content, it poses significant challenges for real-time applications. A sports fan doesn't want to see a game-winning goal on their phone half a minute after it happens in the stadium, and a bettor doesn't want their odds to update seconds behind the action.

Emerging Solutions Compared

Low-Latency HLS: This technology can reduce delays to 2-3 seconds, making it suitable for live sports and events. However, it requires an optimized network infrastructure and support across devices. https://developer.apple.com/documentation/http-live-streaming/

WebRTC: Ideal for peer-to-peer real-time communication, WebRTC delivers near-instantaneous streams. Its scalability, however, is limited, making it less suitable for platforms with millions of concurrent users. https://webrtc.org/

CMAF: Balancing low latency with scalability, CMAF provides an efficient framework for platforms using HLS and DASH. It is particularly effective for platforms with global audiences that demand compatibility and low overhead. https://www.harmonicinc.com/insights/blog/low-latency/

Future Implications

The adoption of CMAF continues to grow as platforms aim to meet the increasing demand for high-quality, low-latency content. By bridging the gap between speed and scalability, CMAF is paving the way for a more responsive and engaging streaming experience. Its applications extend beyond live sports and betting, encompassing e-commerce, virtual events, and interactive media, making it a cornerstone of modern streaming technology.

The Streaming Landscape: An Ever-Evolving Ecosystem

The platforms we use daily are built on a delicate balance of codecs, protocols, and transport mechanisms, each tailored to specific needs:

- **YouTube:** Optimized for scalability and diversity of content types.
- **Netflix:** Prioritizes quality and reliability for cinematic experiences.
- **Twitch:** Balances real-time interactivity with user experience.
- **Zoom:** Sacrifices fidelity for immediacy in communication.

As streaming evolves, so too will these technologies. Emerging trends like edge computing, AI-enhanced delivery, and decentralized platforms promise to further disrupt the landscape. Yet, the core challenge remains: delivering high-quality, real-time content to global audiences at scale.

Streaming's hidden evolution is a story of constant innovation. It's a testament to the engineers and developers who turned once-unimaginable concepts into the seamless experiences we now expect.

Conclusion

The evolution of streaming has been a journey of constant innovation, shaped by the relentless pursuit of quality, efficiency, and interactivity. From the earliest days of RealAudio on dial-up to today's low-latency, 4K, and adaptive streaming experiences, every major leap has been driven by compression breakthroughs, protocol refinements, and smarter delivery networks.

Streaming platforms are no longer just about delivering content. They are about optimizing the experience: balancing real-time responsiveness, scalability, and device compatibility. Each technology stack reflects different priorities:

- **Netflix, Disney+, and Prime Video** optimize for high-quality, on-demand viewing, leveraging DASH, HLS, and TCP to ensure smooth playback.
- **Twitch, YouTube Live, and live sports broadcasters** require low-latency ingestion and playback, carefully balancing UDP,

RTMP, RTP, and CMAF to reduce lag while maintaining stability.

- **Zoom, Discord, and financial trading systems** prioritize instantaneous real-time interaction, relying on Opus, WebRTC, and UDP to keep conversations fluid and responsive.
- **CDNs and Edge Computing** have transformed how content is delivered, moving away from centralized servers to a globally distributed architecture that minimizes latency and maximizes efficiency.

The hardware and infrastructure supporting these experiences have also evolved. What once required expensive, on-premise servers, like the Dell OptiPlex machines we used at Broadcast.com, can now be handled by cloud-based, AI-powered, GPU-accelerated processing or even a consumer laptop.

Yet, despite these advancements, new challenges remain. Latency is still a persistent hurdle for live content, especially in sports, gaming, and remote applications like telemedicine and autonomous vehicles. Compression codecs must keep up with the increasing demand for 4K, 8K, and immersive experiences while balancing the costs of bandwidth, processing power, and licensing fees.

- Looking ahead, the next wave of streaming innovation will likely focus on:
- Smarter adaptive streaming algorithms that adjust in real-time to network conditions, user preferences, and AI-powered optimizations.
- Edge computing and decentralized content delivery to further reduce latency and bandwidth congestion.
- AI-assisted encoding and compression, automating real-time quality enhancements while reducing file sizes without loss of fidelity.
- Next-generation codecs like VVC, EVC, and AV1 drive better compression and efficiency for high-resolution content.
- New protocols are designed for ultra-low-latency experiences, bridging the gap between real-time interactivity and global-scale streaming.

The hidden evolution of streaming technology has shaped the way we consume media today. What was once an experimental, bandwidth-limited endeavor has become the backbone of modern digital entertainment, communication, and commerce.

What do you think will drive the next wave of streaming innovation? Will AI-driven encoding and edge networks finally eliminate buffering? Will WebRTC and CMAF deliver true real-time global streaming?

10: From Broadcast.com to 5G
The Infrastructure That Made Streaming Ubiquitous

Streaming's rise from a mere curiosity to a worldwide staple was never just about cutting-edge codecs, sleek protocols, or clever content delivery networks (CDNs). Sure, those pieces matter, but the real magic lies in the dramatic leaps we've made with broadband and, perhaps more significantly, with mobile networks and the smartphones that ride their signals. Those twin forces built the launching pad for household names like Netflix, YouTube, and Twitch, transforming the once-buffering novelty of online video into a daily ritual for billions. As technology keeps pushing ahead and our appetite for on-demand everything surges, the need for robust infrastructure innovation has never been more urgent.

It's wild to think that in the 1990s, people accessed "streaming" content on landlines via screeching, whirring dial-up modems. It was an agonizingly slow way to get to the early World Wide Web. Today, we have to wonder: how many of us still have a landline at home? Sure, cell phones were around back then, but most people still had an answering machine, a phone on a wire, and maybe even a pager in their pocket. It's a testament to how far we've come that something once so central to internet usage is now little more than a footnote of obsolete techno-archaeology. Meanwhile, an entire generation, let's be honest, Gen Z, has probably never used one.

As a side note, I've still got a little box filled with oddball RJ11 telephone connectors that I had painstakingly collected over the years, souvenirs of the days when I'd pack them for business trips (even international ones) just in case. My son's been gently nudging me to toss them in the pile of outdated tech I finally need to part with, but I'm weirdly attached for now.

A recent study shows that smartphones account for roughly 37% of global online video viewing, compared to about 33% for computers and 29% for smart TVs. This gap reflects our growing tendency to reach for our phones whenever we want to stream, fueled by faster data speeds and our hunger for on-the-go entertainment.

Global online video viewing

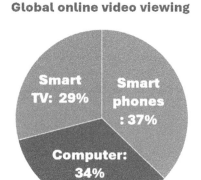

That shift underscores the dominance of mobile devices in our streaming lives, an evolution driven by everything from ever-faster data speeds to our unstoppable craving for convenience and immediacy. https://scoop.market.us/online-video-platform-statistics/

In the pages ahead, we'll unravel how these telecom foundations came to be and why pushing them further is critical right now. We'll dig into the heavyweight players, the cutting-edge technologies, and the social and economic stakes of a future built on streaming. Strap in. It's a story of constant reinvention, where connectivity is king, and the next revolution is always on the horizon.

Deregulation of Communications Companies and the Push to Monetize and Expand Capacity

The deregulation of the telecommunications industry, particularly marked by the Telecommunications Act of 1996, significantly transformed the communications landscape in the United States.

This legislative shift aimed to foster competition by allowing any communications business to enter any market and compete against others, effectively dismantling previous monopolistic structures. https://www.fcc.gov/general/telecommunications-act-1996

One of the primary outcomes of this deregulation was the accelerated monetization and expansion of network capacities. Traditional Plain Old Telephone Service (POTS) lines, which were limited in functionality, began to be overshadowed by more versatile and cost-effective digital infrastructures. The Act encouraged investments in broadband and other advanced services, leading to a significant increase in capital investments by broadband providers. From 1996 through 2014, approximately $1.4 trillion was invested in expanding and upgrading networks. https://www.brookings.edu/articles/was-the-1996-telecommunications-act-successful-in-promoting-competition/

This environment lowered the barriers to entry for nontraditional providers, such as cable companies and internet service providers, enabling them to offer services that were once the domain of established telephony companies. The convergence of services over digital networks meant that companies could now provide a suite of offerings, including voice, data, and video, over a single infrastructure, further promoting competition and innovation. https://en.wikipedia.org/wiki/Network_convergence

However, the anticipated competition in local telephony markets did not fully materialize as expected. Instead, the industry witnessed significant consolidation, with large corporations merging to leverage economies of scale and expand their service offerings. This consolidation sometimes led to reduced competition in certain markets, raising concerns about monopolistic behaviors and the potential negative impacts on consumer prices and service quality. https://www.repository.law.indiana.edu/cgi/viewcontent.cgi?article=1445&context=fclj&httpsredir=1&referer=

Conclusion

The deregulation of communications companies catalyzed a shift toward monetizing and expanding network capacities, offering consumers a broader range of services at potentially lower costs. While it opened the market to nontraditional providers and spurred technological advancements, it also led to industry consolidation, highlighting the complex dynamics between deregulation, competition, and consumer welfare.

The Broadband Revolution: Building the Foundation

In the 1990s, dial-up internet was slow, unreliable, and unsuitable for anything beyond text-based interactions. Streaming audio and video on connections maxing out at 56 Kbps was a frustrating exercise, plagued by buffering and low quality. The broadband revolution, beginning in the late 1990s and early 2000s, changed all that.

Key Developments in Broadband Infrastructure:

In the late 20th century, telecommunications giants such as GTE (now part of Verizon) and AT&T spearheaded the deployment of fiber optic networks, revolutionizing data transmission with unprecedented bandwidth and speed. In 1977, GTE installed one of the first non-experimental fiber optic links, marking a significant milestone in practical telecommunications applications. Race Communications:

https://race.com/blog/history-of-fiber-internet/

Similarly, AT&T conducted early fiber optic field trials in Chicago, further validating the technology's potential.

https://www.ecmag.com/magazine/articles/article-detail/fiber-optic-history-timeline

---

These advancements laid the groundwork for data-intensive applications, including streaming media.

Concurrently, the advent of high-speed internet access technologies like Digital Subscriber Line (DSL) and cable modems transformed consumer connectivity. DSL enabled rapid internet access over existing copper telephone lines, making broadband more affordable and accessible to millions. Cable modems utilized the infrastructure of cable television providers, with companies such as Comcast and Time Warner Cable (now Spectrum) emerging as dominant forces in the broadband market.

Building on these broadband and mobile breakthroughs, Netflix started out as a DVD rental-by-mail company but soon realized streaming held game-changing potential. In 2007, the company introduced an online streaming feature that lets subscribers watch movies and TV shows instantly by taking advantage of the high-speed internet already installed in many homes. This move completely bypassed the standard cable model, delivering on-demand entertainment at the click of a button and helping ignite a wave of "cord-cutting."

Part of cord-cutting's appeal was cost. Cable TV often came bundled with an enormous list of channels that most people never watched, leaving subscribers feeling like they were paying too much for content they didn't use. Remember aimlessly "channel surfing," flipping through one linear channel after another while hoping to find something interesting? Streaming changed the game by letting people pay only for what they wanted and watch whenever they felt like it. In a twist of fate, though, many viewers today end up subscribing to several streaming platforms, sometimes paying as much or more than they did under the old cable regime. Instead of channel surfing, they now face endless scrolling through on-demand menus brimming with suggestions.

In a sense, we've come full circle: the channel lineup has become a stack of apps, and the flow of new shows and movies never seems to stop. Collectively, these developments underscore the critical interplay between technological innovation and the evolution of media consumption, highlighting how advancements in network infrastructure and internet accessibility have paved the way for the streaming-centric landscape we experience today.

Summary:

1. **Fiber Optic Deployment:** Telecommunications giants like **GTE** (now Verizon) and **AT&T** were early leaders in deploying fiber

optic networks. Fiber's unprecedented bandwidth and speed were critical for enabling data-heavy applications like streaming.

2. DSL and Cable Modems
   - **DSL (Asymmetric Digital Subscriber Line):** Allowed high-speed internet access over existing copper telephone lines, bringing affordable broadband to millions.
   - **Cable Modems:** Leveraged cable TV infrastructure, with companies like Comcast and Time Warner Cable (now Spectrum) becoming dominant players.
3. **The Netflix Pivot (2007):** Broadband made Netflix's pivot from DVD rentals to streaming possible. With faster connections, Netflix delivered on-demand movies and TV shows directly to homes, reshaping the entertainment industry and paving the way for the cord-cutting era.

The Influence of Social Networks on Streaming Services

Before social media giants like Facebook, Twitter, and TikTok dominated the online world, early versions of digital communities thrived in the form of Bulletin Board Systems (BBSs), Usenet groups, and online forums. These platforms, dating back to the 1980s and 1990s, allowed users to post messages, share files, and engage in threaded discussions long before the modern news feed existed.

At broadcast.com, we set up forums that enabled real-time discussions and audience interaction, laying the groundwork for how digital communities would later evolve around streaming media. I also helped pioneer one of the earliest interactive forums for authors and readers at Timberwolf Press (2000–2004), where we hosted live author interviews and call-ins from readers, a precursor to the interactive engagement that now defines platforms like Reddit AMAs, Twitter Spaces, and live-streamed Q&As.

As broadband expanded and web technology advanced, these text-heavy discussion boards gradually evolved into the interactive, video-rich social networks we use today. Now, rather than simply discussing media, social platforms have become direct drivers of content discovery, engagement, and even distribution, fundamentally reshaping the way streaming services operate.

Content Discovery and Promotion

Social media platforms have become pivotal in how audiences discover and engage with streaming content. Users frequently encounter recommendations, reviews, and trending topics on platforms like Facebook, Twitter, and TikTok, guiding them toward specific shows, movies, or live events. This organic sharing amplifies visibility and can lead to viral popularity for certain content.

For instance, TikTok has revolutionized the music industry by enabling songs to gain popularity through organic trends or influencer-driven campaigns. The platform's global reach and algorithm-driven content delivery have made it a significant promotional tool for artists and record labels. https://www.businessinsider.com/how-tiktok-is-changing-music-industry

Live Streaming and Real-Time Engagement

The rise of live streaming on social media has reshaped how audiences consume events, news, and entertainment. Platforms such as Facebook Live, Instagram Live, and Twitch offer real-time broadcasting, allowing content creators and brands to interact directly with viewers. This immediacy fosters a sense of community and enhances user engagement.

For example, live streaming has transformed social media marketing by enabling businesses to showcase products and engage with their audience in real-time, leading to increased customer loyalty and brand visibility. https://www.nobledesktop.com/learn/social-media-marketing/the-impact-of-live-streaming-on-social-media-marketing

User-Generated Content and Community Building

Social networks empower users to create and share their own content related to streaming services. Fan-made reviews, reaction videos, and discussion threads contribute to a communal viewing experience, even when audiences are geographically dispersed. This user-generated content fosters deeper connections among viewers and between audiences and creators.

For instance, platforms like Twitch have cultivated communities where viewers and streamers interact closely, creating a participatory culture around gaming and other live-streamed content.

https://kadence.com/en-us/how-live-streaming-is-reshaping-media-consumption

Influence on Content Creation and Curation

The feedback loop between social media and streaming services influences the types of content produced. Creators monitor social media trends and audience preferences to tailor their offerings, ensuring relevance and appeal. Streaming platforms also utilize data analytics from social interactions to curate personalized recommendations, enhancing user satisfaction and retention.

For example, social media and creators drive viewers to TV shows, movies, and games, with more people turning to social media to discover new and exciting content. https://www2.deloitte.com/us/en/insights/industry/technology/digital-media-trends-consumption-habits-survey/2024/online-creators-and-the-impact-of-social-media-on-entertainment.html

Challenges and Considerations

While the convergence of social networks and streaming services offers numerous benefits, it also presents challenges. Issues such as content moderation, digital piracy, and the spread of misinformation require ongoing attention. Additionally, the pressure on creators to produce content that aligns with rapidly changing social media trends can impact the quality and diversity of streaming offerings.

In summary, the interplay between social networks and streaming services has revolutionized the media landscape, creating dynamic platforms for content discovery, engagement, and creation. As these technologies continue to evolve, they will undoubtedly further shape the ways in which audiences consume and interact with media.

Apple's Influence on Streaming Services and Digital Media Distribution

Despite Apple's undeniable success, the initial iPhone wasn't *completely* successful with *every single* user. I was in a particular founder's truck when, after repeated problems and frustrations, he threw his phone out the window of his speeding vehicle. It *was,* after all, a dramatic departure from what most people were used to.

Despite this now amusing memory, Apple has certainly played a pivotal role in shaping the landscape of streaming services and digital media distribution through strategic initiatives and innovative platforms.

Apple TV+: A Foray into Original Content

Launched in November 2019, Apple TV+ marked Apple's entry into the competitive streaming market, focusing on original programming. Despite its high-quality content, the service has faced challenges in gaining significant market share, holding approximately **11% of the American streaming market** as of early 2025. To enhance accessibility and attract a broader audience, Apple has expanded the availability of Apple TV+ to Android devices, allowing users of various platforms to access its content. https://www.sfgate.com/streaming/article/apple-tv-plus-free-weekend-20013725.php, https://www.theaustralian.com.au/business/technology/apple-tv-finally-comes-to-android-phones-as-the-tech-titan-woos-new-customers/news-story/cf043a11bc9d5328560b183c5e9ff45a

Apple Music: Transforming Music Consumption

Apple Music, introduced in 2015, has significantly impacted the music industry by providing a new revenue stream and democratizing access to music for both major labels and independent artists. However, it has also introduced complex royalty models that often leave artists with less income compared to **traditional sales, prompting ongoing adjustments in business models and revenue distribution.** https://aimm. edu/blog/how-streaming-changed-the-music-industry

App Store Ecosystem: Facilitating Digital Media Distribution

The Apple App Store has been instrumental in the distribution of digital media, including streaming services. It has enabled developers to reach a global audience, offering a platform for both large-scale streaming services and niche content providers. The App Store's ecosystem has supported the growth of various entertainment services, including live-streaming apps, contributing to the diversification of available content. https://www.apple.com/newsroom/pdfs/the-continued-growth-and-resilience-of-apples-app-store-ecosystem.pdf

Challenges and Strategic Adjustments

Despite its successes, Apple has faced challenges in the streaming domain. Apple TV+ has struggled to compete with established giants like Netflix, which holds a 55% market share. To address this, Apple has offered free trial weekends and expanded its service to non-Apple devices, aiming to increase its subscriber base. https://www.sfgate.com/streaming/article/apple-tv-plus-free-weekend-20013725.php

Conclusion

Apple's strategic initiatives in streaming services and digital media distribution have significantly influenced the industry. Through platforms like Apple TV+ and Apple Music, the company has expanded its ecosystem, offering diverse content and services. While facing challenges in a competitive market, Apple's continuous adaptation and innovation underscore its commitment to shaping the future of digital media consumption.

SMS Texting and Alternative Messaging Platforms: Their Impact on Streaming Services

The evolution of communication from traditional Short Message Service (SMS) to alternative messaging platforms has significantly influenced the landscape of streaming services. These changes have affected how users discover, engage with, and share streaming content, creating new opportunities for content distribution and marketing.

Transition from SMS to OTT Messaging

Initially, SMS was the primary method for mobile text communication, limited to 160-character messages and subject to carrier fees. With technological advancements, Over-The-Top (OTT) messaging services like WhatsApp, iMessage, Telegram, and Facebook Messenger emerged, offering rich media sharing, group chats, and enhanced interactivity. These platforms operate over the Internet, bypassing traditional cellular networks and providing cost-effective communication options for both users and businesses.https://en.wikipedia.org/wiki/Over-the-top_media_service

Influence on Content Discovery and Sharing

OTT messaging platforms have transformed how users discover and share streaming content. Features such as in-app video sharing, direct links, and group discussions enable seamless content recommendations, facilitating organic growth for streaming platforms.

For instance, services like WhatsApp allow users to share Netflix show recommendations or YouTube videos directly within a chat, prompting immediate viewing or discussion. The ease of one-on-one and group sharing creates viral marketing opportunities that drive engagement without requiring formal advertising campaigns.

Integration with Streaming Services

Streaming platforms have recognized the potential of integrating with messaging apps to enhance user experience. Many services now allow users to share content directly from within their apps, and some have experimented with co-viewing features where users can watch together while chatting in real-time.

Examples include:

- **Netflix's "Watch Party" features** which allow synchronized viewing across different locations.
- **Twitch's chat integration**, where live audiences engage with streamers through real-time messaging.
- **Apple's SharePlay** which lets users watch streaming content together over FaceTime.

Marketing and User Engagement

The shift to OTT messaging has also opened new avenues for direct marketing within streaming services. Platforms now utilize messaging apps for:

- Personalized content recommendations
- Exclusive promotions and subscriber deals
- Alerts for new episodes or live events

OTT messaging campaigns often achieve higher engagement rates than email marketing, as users are more likely to open a direct message than a promotional email. https://www.textmagic. com/blog/10-online-text-messaging-advantages-you-havent-considered

Challenges and Considerations

While OTT messaging platforms offer numerous benefits, they also present challenges, including:

- **Content moderation concerns**: Private messaging makes it harder to regulate piracy and misinformation.
- **Data privacy issues:** Many platforms encrypt messages, limiting how companies can track user interactions.
- **Platform dependency**: Streaming services integrating with third-party messaging apps risk losing control over how content is shared.

Conclusion

The evolution from SMS to OTT messaging platforms has significantly impacted streaming services by:

- Altering content discovery methods
- Enhancing user engagement through real-time interactions
- Providing new marketing and promotional tools

As these technologies continue to evolve, their integration with streaming platforms is expected to deepen, further transforming the digital media landscape.

The Influence of the Adult Entertainment Industry on Streaming Services

The adult entertainment industry has been a significant driver in the adoption and advancement of streaming technologies, shaping the broader digital media landscape.

Early Adoption of Streaming Technologies

In the 1990s, the adult industry was among the first to embrace online streaming, recognizing its potential to deliver content directly to consumers. Indeed, the industry took to streaming like a duck to water.

Like it or not, it undeniably played a crucial role in popularizing streaming media and demonstrated the viability of online content distribution.

Advancements in Payment Processing and Security

The need to monetize adult content led to innovations in secure online payment systems. The industry's demand for discreet and efficient billing solutions contributed to the development of payment processing technologies that are now standard across various e-commerce platforms.

Influence on Content Delivery Networks (CDNs)

To efficiently deliver high-bandwidth content to a global audience, the adult industry invested in and promoted the development of Content Delivery Networks (CDNs). These networks have since become essential infrastructure for streaming services, ensuring fast and reliable content delivery.

Pioneering User Engagement and Personalization

The adult industry has been at the forefront of implementing user engagement strategies, such as personalized recommendations and interactive features. These innovations have been adopted by mainstream streaming services to enhance user experience and retention.

Driving Bandwidth and Compression Innovations

The high demand for adult content has pushed the limits of internet bandwidth, prompting advancements in data compression and streaming technologies. These improvements have benefited the entire streaming industry by enabling higher-quality video delivery over existing infrastructure.

Conclusion

The adult entertainment industry's early and enthusiastic adoption of streaming technologies has had a lasting impact on the development of digital media. Its contributions to payment processing, content delivery, user engagement, and technological innovation have helped shape the streaming services we use today.

eBay's Influence on Streaming Services and Digital Media

eBay, established in 1995 as one of the first successful online auction platforms, played a significant role in shaping the digital landscape. While not directly a streaming service, eBay's influence can be seen in user engagement models, digital media distribution, and the integration of streaming into e-commerce strategies.

Pioneering Online Platforms and User Engagement

As an early internet marketplace, eBay demonstrated the power of user-driven content and peer-to-peer transactions, setting the stage for platforms that rely on user engagement, such as YouTube, Twitch, and modern streaming services. eBay helped popularize online reviews, seller ratings, and recommendation algorithms, key elements now found in streaming platforms to personalize content and build community trust.   https://www.axios.com/2018/06/24/ebay-shaped-modern-internet-google-facebook

Integration of Streaming Technologies for Internal Communication

eBay has used streaming technologies internally to enhance communication and collaboration. The company implemented an enterprise YouTube-style platform, allowing employees to upload and share videos for training, meetings, and global teamwork. This mirrors how many companies now use streaming as a primary mode of communication, extending beyond entertainment into professional and educational spaces. https://www.streamingmedia.com/Articles/Editorial/Featured-Articles/eBay-Embraces-Enterprise-YouTube-97146.aspx

Influence on Digital Media Players and Content Access

The rise of digital media players, such as Roku, Fire TV, and Android TV boxes, has been influenced by eBay and other online marketplaces. eBay provided a platform for selling media devices, including early MP3 players, DVD players, and, later, streaming devices, making digital content more accessible to a broader audience. However, eBay has also been a hub for gray-market streaming devices, where users purchase pre-configured hardware that enables access to unauthorized content, raising legal concerns within the industry. https://en.wikipedia.org/wiki/Digital_media_player

Collaborations and Sponsorships in the Streaming Era

eBay has expanded into digital media through strategic sponsorships and integrations with streaming content. One example is its collaboration with the hit TV show Love Island, where eBay promoted sustainable fashion through its "Pre-Loved Island" campaign. This partnership leveraged streaming audiences and demonstrated how e-commerce and streaming can intersect in modern marketing strategies. https://www.thecurrent.com/tv-ctv-ebay-streaming-soulmate-love-island

Conclusion

eBay's contributions to digital engagement, streaming adoption in workplaces, accessibility of media devices, and integration with modern streaming content have made it a key player in the evolution of digital media. While primarily an e-commerce platform, its innovations in user interaction, digital transactions, and content promotion have helped shape the online media landscape that streaming services now thrive within.

Content Licensing: Owning vs. Leasing

In the days of VHS tapes and DVDs, buying a movie meant physically owning that copy. You could lend it to friends, sell it, or simply let it collect dust on your shelf. With digital delivery and streaming services, however, many people discover they never truly own the content they have paid for. Instead, they hold a license that can change or be revoked at any time.

A memorable example occurred in 2009, when Amazon remotely deleted certain Kindle copies of George Orwell's "1984," stirring public outcry over digital ownership. https://www.nytimes.com/2009/07/18/technology/companies/18amazon.html

In a similar vein, Amazon attracted attention again in 2020 for arguing in court that customers do not genuinely own the digital movies and shows they purchase on Prime Video. https://news.bloomberglaw.com/litigation/amazon-cant-dislodge-purchased-digital-content-deception-suit

As a result, titles on streaming platforms may be removed or edited if a licensing agreement expires or changes, leaving consumers with few options. This stands in sharp contrast to the era of physical media, where the copy you bought was yours to keep, unaltered.

More recently, Amazon removed the "Download & Transfer via USB" option for Kindle users, eliminating their ability to locally backup or transfer their purchased e-books outside Amazon's ecosystem. Now, Kindle users can only access their purchased books through Amazon's cloud services, further reinforcing the idea that they are leasing, rather than owning, their digital content. https://www.theverge.com/news/612898/amazon-removing-kindle-book-download-transfer-usb

These cases highlight a growing shift in the way media is controlled. When you purchase a physical disc, that copy is yours indefinitely, free from updates, removals, or restrictions. In contrast, digital content often comes with hidden strings, whether it is a movie that disappears from a library, a book that can no longer be transferred, or a streaming service that alters content after the fact. As platforms tighten their grip on digital distribution, true ownership of media is becoming harder to maintain.

The Mobile Revolution: Streaming on the Go

If broadband brought streaming to homes, mobile networks made it ubiquitous. The leap from 3G to 5G turned smartphones into powerful streaming hubs, enabling users to watch, listen, and interact anytime, anywhere.

Key Milestones in Mobile Connectivity:

The evolution of mobile connectivity has been marked by significant milestones, each contributing to the enhancement of streaming capabilities and reshaping how content is consumed on the go.

3G Networks: The Starting Point

Introduced commercially in 2001, 3G networks marked a pivotal shift from voice-only communications to data-centric services. With data transfer rates reaching up to 7.2 Mbps, 3G enabled basic mobile internet access, including web browsing and initial forms of video streaming. This development laid the groundwork for mobile video consumption, albeit with limitations in speed and quality. https://www.forbes.com/councils/forbestechcouncil/2020/12/18/3g-4g-5g-and-beyond-the-quest-for-mobile-connectivity-and-speed/

4G LTE: Streaming Goes Mobile

The advent of 4G Long-Term Evolution (LTE) technology brought a substantial increase in data transmission speeds, reaching up to 100 Mbps. This leap facilitated seamless high-definition video streaming on mobile devices, propelling the popularity of platforms like Twitch and TikTok. Consequently, smartphones became the primary devices for streaming content, offering users the flexibility to consume media anytime and anywhere... https://www.qualcomm.com/5g/what-is-5g

5G: The Future of Streaming

The emergence of 5G technology promises to revolutionize mobile connectivity with speeds up to 20 Gbps and ultra-low latency. These advancements open avenues for real-time sports streaming, augmented and virtual reality experiences, and cloud gaming. Additionally, 5G enables features like geo-fencing and localized content delivery, enhancing in-stadium sports experiences by providing spectators with personalized, real-time information and services. https://www.qualcomm.com/5g/what-is-5g

Each generation of mobile connectivity has progressively enhanced the streaming experience, transforming mobile devices into powerful platforms for consuming rich, data-intensive content.

Summary:

1. **3G Networks: The Starting Point:** Launched in the early 2000s, 3G introduced mobile internet capable of basic streaming. While limited in speed, it marked the beginning of mobile video consumption.
2. **4G LTE: Streaming Goes Mobile:** With speeds up to 100 Mbps, 4G LTE enabled seamless mobile streaming, fueling the growth of platforms like Twitch and TikTok. Smartphones have become the primary streaming devices for millions worldwide.
3. **5G: The Future of Streaming:** Offering speeds up to 10 Gbps and ultra-low latency, 5G has unlocked possibilities like

real-time sports streaming, AR/VR experiences, and cloud gaming. Geo-fencing and localized content delivery have also become critical for in-stadium sports experiences.

Wi-Fi: Expanding Broadband Indoors

While mobile networks took streaming outdoors, Wi-Fi ensured seamless connectivity indoors and in public spaces.

Key Advancements in Wi-Fi:

The progression of Wi-Fi technology has significantly enhanced streaming capabilities and integrated connectivity into daily life.

Evolution of Wi-Fi Standards

The development of Wi-Fi standards has been pivotal in improving wireless communication. Introduced in 2009, Wi-Fi 4 (802.11n) offered speeds up to 600 Mbps, operating on both 2.4 GHz and 5 GHz bands. This advancement facilitated more reliable connections and supported high-definition streaming. Building upon this, Wi-Fi 6 (802.11ax) and its extension, Wi-Fi 6E, brought multi-gigabit speeds and reduced network congestion, even in high-density environments. These improvements have enabled seamless 4K streaming and more efficient performance in smart homes populated with numerous connected devices. https://www.cbtnuggets. com/blog/technology/networking/when-to-use-802-11-a-b-g-b-nc-wifi-standards

Smart Home Integration

Wi-Fi has become the backbone of smart home ecosystems, powering devices such as smart TVs, streaming gadgets, and voice assistants. This integration has embedded streaming into everyday life, allowing users to control entertainment systems, lighting, and security features through centralized, Wi-Fi-enabled platforms. The emergence of standards like Matter aims to simplify and secure smart home device interoperability across different brands and ecosystems, further enhancing the seamless integration of streaming and other services into daily routines. https://www.wired.com/story/what-is-matter/

Public Wi-Fi Expansion

The widespread availability of public Wi-Fi in venues like cafes, airports, and stadiums has extended the reach of streaming services, making mobile viewing a routine activity. This accessibility allows individuals to enjoy their favorite content on the go, transforming how and where media is consumed. Innovations such as community-wide managed Wi-Fi solutions, like Calix SmartTown, provide seamless connectivity in public spaces, further embedding streaming into the fabric of daily life. https://www.calix.com/products/smartlife/smarttown.html

Collectively, these advancements in Wi-Fi technology have not only enhanced streaming quality and accessibility but have also woven connectivity into the everyday experiences of users worldwide.

Summary:

1. Evolution of Wi-Fi Standards:
   - **Wi-Fi 4 (802.11n):** Speeds up to 600 Mbps.
   - **Wi-Fi 6 and 6E:** Introduced multi-gigabit speeds and reduced congestion, enabling 4K streaming in high-density environments.
2. **Smart Home Integration:** Wi-Fi powers smart TVs, streaming devices, and voice assistants, embedding streaming into everyday life.
3. **Public Wi-Fi:** Available in cafes, airports, and stadiums, public Wi-Fi extended streaming's reach, making mobile viewing a daily activity.

Why Infrastructure Matters Now

Streaming infrastructure has consistently driven innovation, and its significance has only intensified in today's interconnected world. Beyond entertainment, streaming now underpins essential sectors such as work, education, and healthcare. Platforms like Zoom and Microsoft Teams facilitate remote work and virtual classrooms, while telemedicine services enable healthcare delivery, all relying on robust streaming capabilities. As these real-time applications expand, the demand for reliable and efficient infrastructure becomes increasingly critical.

Emerging technologies, including live sports betting, cloud gaming, and augmented and virtual reality (AR/VR), require ultra-low latency and high-speed networks to function effectively. Advancements in broadband, 5G, and Wi-Fi 6E are pivotal in supporting these applications, ensuring seamless and responsive user experiences. For instance, the deployment of 5G networks has been instrumental in enhancing streaming quality and reducing latency, thereby enabling more immersive and interactive applications. https://www.yellowbrick.co/blog/entertainment/the-future-of-streaming-technology-key-trends-and-innovations

Despite these technological advancements, a significant portion of the global population still lacks access to high-speed internet, limiting their participation in the digital economy. Initiatives like Starlink and other satellite-based systems aim to bridge this digital divide by bringing broadband connectivity to rural and underserved areas, thereby expanding access to streaming services and other online resources.

The proliferation of high-definition formats such as 4K and 8K, along with interactive streaming, has led to a surge in data consumption, pushing existing bandwidth limits. Innovations in edge computing and AI-driven optimizations are essential for managing this data explosion, reducing costs, and minimizing environmental impacts. By processing data closer to the source and utilizing AI to optimize streaming workflows, these technologies contribute to more sustainable and efficient streaming infrastructures. https://www.harmonicinc.com/insights/blog/video-trends/

Streaming has become a significant economic driver, creating jobs and supporting creative industries worldwide. For instance, Netflix's investment in South Korea since 2016 has generated almost ₩5.6 trillion (approximately \$4.7 billion) in economic impact and created more than 16,000 jobs. https://www.fdiintelligence.com/content/locations/global/fresh-content-how-streaming-platforms-are-changing-the-geography-of-media-investment-80490

In the United States, digital music streaming supported over 92,000 total jobs in 2021, reflecting its substantial contribution to the economy. https://dima.org/news-and-resources/new-study-highlights-music-streaming-outsized-impact-us-economy/

The continued expansion of streaming services necessitates robust infrastructure to meet global demand and ensure equitable access. Platforms like Netflix, YouTube, and Twitch leverage the latest infrastructure advancements to enhance user experience, reduce subscriber churn, and maintain a competitive edge in the rapidly evolving digital landscape. Technological advancements such as 5G connectivity and edge computing facilitate faster streaming speeds and reduced latency, enabling platforms to deliver immersive experiences and cater to diverse viewer preferences. https://www.muvi.com/blogs/the-technological-advancements-in-ott-platforms/

Moreover, the integration of artificial intelligence (AI) in streaming platforms has reshaped the industry by enhancing user engagement and driving business success. AI-powered recommendations, adaptive streaming, interactive content, and multilingual support have become essential features that improve viewer loyalty and contribute to product growth. https://inoxoft.com/blog/essential-features-for-streaming-platforms/

As the streaming industry continues to evolve, platforms must adapt to changing viewer habits and technological advancements to remain relevant. This includes investing in infrastructure that supports high-quality streaming, implementing user-friendly interfaces, and offering personalized content to meet the expectations of a global audience. By doing so, streaming services can sustain their growth and continue to play a pivotal role in the entertainment ecosystem.

In summary, the importance of streaming infrastructure today extends far beyond entertainment, encompassing critical aspects of daily life, economic development, and technological innovation.

Summary:

1. **Global Dependence on Streaming:** Streaming isn't just about entertainment. It powers work (Zoom, Microsoft Teams), education (online classes), and healthcare (telemedicine).

As real-time applications expand, the demand for reliable infrastructure grows.

2. **Real-Time Applications:** Emerging technologies like live sports betting, cloud gaming, AR/VR, and interactive streaming require ultra-low latency and high-speed networks. Advancements in broadband, 5G, and Wi-Fi 6E are critical to enabling these experiences.

3. **Bridging the Digital Divide:** Billions of people still lack access to high-speed internet, limiting their participation in the digital economy. Technologies like Starlink and other satellite systems are crucial for bringing broadband to rural and underserved areas.

4. **Data Explosion and Sustainability:** With 4K, 8K, and interactive formats pushing bandwidth limits, innovations in edge computing and AI-driven optimizations are essential for reducing costs and minimizing environmental impacts.

5. **Economic and Cultural Impact:** Streaming drives significant economic activity, creating jobs and supporting creative industries. Its expansion requires robust infrastructure to support global demand and ensure equitable access.

6. **Platform Competition:** For platforms like Netflix, YouTube, and Twitch, leveraging the latest infrastructure advancements enhances user experience, reduces churn, and strengthens their competitive edge.

Key Players in the Infrastructure Revolution

GTE/Verizon

GTE, which merged with Bell Atlantic to form Verizon Communications in 2000, was a pioneer in deploying fiber optic networks during the 1990s. Recognizing the limitations of traditional copper lines in meeting future data demands, GTE invested heavily in fiber optics to enhance data transmission capabilities. Following the merger, Verizon continued this trajectory by launching its FiOS (Fiber Optic Service) network in 2005, offering consumers high-speed internet and television services over a fiber-optic infrastructure. https:// www.verizon.com/about/news/press-releases/two-years-after-launch-verizon-fios-tv- continues-break-new-ground-subscription-tv-market

Today, Verizon remains at the forefront of broadband innovation with its 5G Ultra Wideband network, which complements its existing fiber backbone. The integration of 5G technology with fiber infrastructure creates a seamless ecosystem capable of supporting data-intensive applications, including high-definition streaming and real-time communications. https://www.verizon.com/about/our-company/ high-speed-broadband

Looking ahead, Verizon is poised to play a significant role in the advancement of edge computing. By processing data closer to the user, edge computing reduces latency, enhancing the performance of applications such as cloud gaming and real-time streaming. Verizon's ongoing investments in both 5G and edge computing technologies position the company to meet the evolving demands of consumers and businesses in an increasingly connected world.

Summary:

Historical Role: GTE was among the earliest companies to invest heavily in fiber optic networks in the 1990s, recognizing that traditional copper lines couldn't meet future data demands. Following the merger with Bell Atlantic in 2000, GTE became Verizon, a leader in broadband innovation. Verizon's FiOS network, launched in 2005, was one of the first widespread consumer fiber-optic services, offering unparalleled speed and reliability at the time.

Relevance Today: Verizon continues to drive innovation with its 5G Ultra Wideband network, which complements its fiber backbone. The integration of 5G with broadband fiber creates a seamless ecosystem for streaming and other data-intensive applications.

Future Outlook: Verizon is expected to play a significant role in edge computing, where content and data processing occur closer to the user, reducing latency for applications like cloud gaming and real-time streaming.

AT&T

AT&T has played a pivotal role in expanding broadband access since the late 1990s, employing a combination of Digital Subscriber Line (DSL) and fiber-optic deployments. This hybrid strategy enabled the company to extend high-speed internet services to both urban and rural areas, thereby connecting a broader range of households to faster internet. In the early 2000s, AT&T's DSL services flourished as fixed-broadband competitors to cable, contributing to a virtuous cycle of improvement and enhancement in the U.S. broadband landscape. https://progressivepolicy.org/wp-content/uploads/2014/03/2014.03-Ehrlich_A-Brief-History-of-Internet-Regulation.pdf

Today, AT&T stands as a major player in both fiber and wireless connectivity, supporting the increasing demands of streaming platforms, smart homes, and connected devices. The company's strategic acquisitions, such as those of WarnerMedia, underscore its focus on integrating content creation with distribution, enhancing its offerings in the entertainment sector. Additionally, AT&T's recent launch of an industry-first integrated fiber and 5G gateway for business customers exemplifies its commitment to innovation in connectivity solutions. https://about.att.com/story/2024/integrated-gateway.html

Looking ahead, AT&T is poised to expand its 5G and fiber networks further, aiming to provide the infrastructure necessary for immersive streaming experiences, including augmented reality (AR) and virtual reality (VR). The company's initiatives in edge computing and low-latency networks are expected to support these advanced applications, ensuring seamless and responsive user experiences. https://about.att.com/blogs/2023/upgrading-the-network.html

Summary:

Historical Role: AT&T was instrumental in expanding broadband access in the late 1990s and early 2000s through a combination of DSL and fiber-optic deployments. Its commitment to hybrid strategies allowed it to reach urban and rural areas, helping more households access faster internet.

Relevance Today: AT&T has evolved into a major player in fiber and wireless connectivity, supporting the demands of streaming platforms, smart homes, and connected devices. Its strategic acquisitions, including WarnerMedia, highlight its focus on merging content creation with distribution.

Future Outlook: As AT&T continues expanding its 5G and fiber networks, its role in providing infrastructure for immersive streaming experiences, like augmented reality (AR) and virtual reality (VR), will grow.

Comcast and Time Warner Cable

In the late 1990s and early 2000s, cable companies such as Comcast and Time Warner Cable (now part of Charter Communications under the Spectrum brand) transformed the broadband landscape by utilizing their existing coaxial cable infrastructures to deliver high-speed internet alongside television services. This approach allowed them to offer internet speeds that significantly surpassed those of Digital Subscriber Line (DSL) services at the time, positioning them as early leaders in the broadband market. https://en.wikipedia.org/wiki/Time_Warner_Cable

Today, these companies continue to dominate the provision of high-speed broadband in urban and suburban areas. Comcast's Xfinity and Spectrum Internet services offer gigabit-speed plans designed to support households with multiple devices streaming simultaneously, catering to the growing demand for seamless streaming experiences.

Looking ahead, cable operators face increasing competition from fiber and satellite internet providers. However, advancements like DOCSIS 4.0 (Data Over Cable Service Interface Specification) aim to keep coaxial networks competitive by enabling symmetrical multi-gigabit speeds and reduced latency. By integrating their broadband services with streaming platforms and smart home ecosystems, these companies strive to maintain and grow their market share in the evolving digital landscape. https://www.cablelabs.com/technologies/docsis-4-0-technology

In summary, Comcast and Time Warner Cable have played pivotal roles in advancing broadband technology, leveraging their existing infrastructures to meet the increasing demands of consumers and remain competitive in the rapidly evolving internet service market.

Summary:

Historical Role: Cable companies revolutionized broadband by leveraging their coaxial cable infrastructure to deliver high-speed internet alongside TV services. Comcast and Time Warner Cable (now Spectrum) were early leaders, offering speeds that far outpaced DSL during the late 1990s and early 2000s.

Relevance Today: These companies remain dominant in providing high-speed broadband in urban and suburban areas. Comcast's Xfinity and Spectrum Internet continue to support streaming services with gigabit-speed offerings tailored for households with multiple devices streaming simultaneously.

Future Outlook: Cable companies face competition from fiber and satellite providers, but innovations like DOCSIS 4.0 could keep coaxial networks competitive. By integrating their broadband services with streaming platforms and smart home ecosystems, these companies will aim to retain their market share.

Qualcomm

Qualcomm has been instrumental in advancing mobile connectivity technologies, notably through the development of Long-Term Evolution (LTE). LTE has become one of the fastest-growing wireless technologies, delivering faster and better mobile broadband experiences on a global scale. https://www.qualcomm.com/research/5g/4g

Today, Qualcomm continues to lead in the development of 5G chipsets, enabling smartphones, Internet of Things (IoT) devices, and even autonomous vehicles to operate on high-speed networks. The company's Snapdragon processors are foundational to mobile devices worldwide, providing robust performance and connectivity. https://www.qualcomm.com/products/technology/5g

Looking ahead, as streaming evolves toward augmented reality (AR), virtual reality (VR), and real-time interactive applications, Qualcomm's advancements in ultra-low-latency 5G and AI-powered processors will be critical. The company is also expected to play a role in next-generation wearables and connected devices designed for streaming-rich environments. https://www.qualcomm.com/products/technology/5g

Summary:

Historical Role: Qualcomm played a pivotal role in developing mobile connectivity technologies, particularly with the introduction of LTE (Long-Term Evolution). By enabling high-speed data transfer on smartphones, Qualcomm helped unlock the potential of mobile streaming, propelling platforms like YouTube, Twitch, and TikTok

Relevance Today: Qualcomm continues to lead in the development of 5G chipsets, enabling smartphones, IoT devices, and even autonomous vehicles to operate on high-speed networks. Its Snapdragon processors are foundational to mobile devices worldwide

Future Outlook: As streaming evolves toward AR/VR and real-time interactive applications, Qualcomm's advancements in ultra-low-latency 5G and AI-powered processors will be critical. The company is also expected to play a role in next-generation wearables and connected devices designed for streaming-rich environments.

Starlink

Starlink, developed by SpaceX, has rapidly transformed global broadband access since its first satellite launches in 2019. Aiming to provide high-speed, low-latency internet to underserved and remote areas worldwide, Starlink surpassed 1 million subscribers by December 2022 and reached 4 million by September 2024. (https://en.wikipedia.org/wiki/Starlink)

Relevance Today

Bridging the Digital Divide: Starlink is crucial for regions lacking traditional broadband infrastructure, such as rural communities, developing nations, and areas affected by natural disasters. By offering connectivity where fiber and cable cannot reach, Starlink is expanding the potential audience for streaming platforms globally.

Support for Streaming: While early satellite internet services suffered from high latency, Starlink's low-earth-orbit (LEO) satellites drastically reduce delays, making it suitable for streaming HD and 4K content.

Applications Beyond Streaming: Beyond consumer broadband, Starlink is enabling remote education, telemedicine, and real-time communication in remote industries like shipping, aviation, and energy exploration.

Future Outlook

Expansion and Scalability: Starlink plans to expand its satellite constellation to over 40,000 units, further improving coverage and capacity. This could revolutionize connectivity for remote regions and provide an alternative to traditional broadband in urban markets.

Market Competition: Starlink is paving the way ahead of competitors like Amazon's Project Kuiper and OneWeb, which are also pursuing LEO satellite-based internet. This competition, should it mature, will likely drive innovation and lower costs, making satellite broadband more accessible.

Integration with 5G and IoT: As Starlink integrates with terrestrial 5G networks and supports IoT devices, it could become a cornerstone of hybrid connectivity solutions. This will be especially critical for industries and regions requiring reliable, global connectivity.

SpaceX and T-Mobile

In August 2022, T-Mobile and SpaceX announced a partnership to eliminate cellular dead zones across the United States by integrating Starlink's satellite technology with T-Mobile's mid-band spectrum. This collaboration aims to create a new network that broadcasts directly from Starlink's satellites, providing near-complete coverage in areas previously lacking cellular service. (https://www.t-mobile.com/news/un-carrier/t-mobile-takes-coverage-above-and-beyond-with-spacex)

The Federal Communications Commission (FCC) approved this initiative in November 2024, marking the first collaboration between a satellite operator and a wireless carrier authorized to provide supplemental coverage from space. The service is expected to begin with text messaging, with plans to expand to voice and data services in the future. https://www.theverge.com/2024/11/27/24307394/starlink-spacex-tmobile-direct-to-cell-satellite-fcc-approval

This partnership not only enhances connectivity in remote and underserved areas but also has significant implications for streaming services. By leveraging Starlink's low-earth-orbit satellites, the collaboration aims to reduce latency and improve data speeds, making high-quality streaming more accessible to a broader audience. As the service evolves to include voice and data capabilities, it holds the potential to further integrate streaming into daily life, regardless of location.

In summary, the T-Mobile and Starlink partnership represents a significant advancement in bridging connectivity gaps, with promising prospects for the future of streaming and real-time applications.

Why This Matters Now

The roles of these key players underscore how broadband and mobile networks have driven the streaming revolution while highlighting the urgent need for continued innovation. As demand for high-quality, real-time, and accessible streaming grows, the infrastructure they provide will be foundational to addressing challenges and unlocking opportunities.

The Road Ahead

The story of streaming's explosive growth is inseparable from the infrastructure that made it possible. From the early broadband deployments of GTE and AT&T to the seamless mobile experiences enabled by Qualcomm's LTE and 5G chipsets, each leap in connectivity spurred an equally important leap in how we consume content. Cable operators like Comcast and Time Warner Cable leveraged existing coaxial networks to bring high-speed internet to homes, Wi-Fi standards evolved to power streaming on every screen, and satellite innovators like Starlink bridged gaps where other infrastructures couldn't reach. Even newer partnerships, such as T-Mobile and SpaceX collaborating to eliminate cellular dead zones, highlight how merging technologies can further push the boundaries of real-time connectivity.

Where we stand now is at the threshold of a new era in which low-latency, high-bandwidth networks are no longer optional luxuries but essential foundations of modern life. Work, education, healthcare, entertainment, and immersive experiences like AR/VR all depend on advanced streaming infrastructure to function efficiently and equitably. The collective challenge moving forward is to extend these innovations to everyone, closing the digital divide, ensuring sustainability, and continuing to refine new methods like edge computing and AI-driven content delivery.

Streaming is no longer just entertainment; it is a pillar of digital transformation, shaping how we work, communicate, and experience the world. As infrastructure continues to evolve, so will our ability to experience content ubiquitously and interactively, ushering in the next wave of transformative streaming applications across the globe.

What's Your Take?

Streaming has come a long way, from buffering videos on dial-up connections to high-definition, real-time content that is accessible anywhere. But where does it go from here? How do you see broadband, mobile networks, and new technologies shaping the next phase of digital media?

11: From Broadcast.com to Today

Why Live Streaming Still Faces a 30-Second Delay

In the 1990s, streaming technology primarily relied on packet-based methods, most notably UDP (User Datagram Protocol), and, to a lesser extent – though certainly not from a lack of ambition – multicast streaming. Both approaches involved breaking down an audio or video signal into small segments known as packets, which were then transmitted individually to the receiving server or end-user.

Due to the unpredictable nature of the early internet infrastructure, these packets often took different routes, occasionally arriving out of sequence, delayed, or sometimes not arriving at all. The receiving system then faced the challenging task of piecing Humpty-Dumpty back together again. This complicated process of reassembling and synchronizing the packets inevitably introduced latency into the playback experience, frequently causing delays of 30 seconds or often longer, thus significantly impacting the immediacy and real-time responsiveness of streaming media.

If UDP was blocked by a firewall or network configuration (common in corporate or restrictive environments), RealNetworks' RealPlayer would revert to TCP streaming, usually on port 80 (HTTP), known as HTTP tunneling. Although TCP guaranteed packet delivery, this introduced additional overhead and increased latency even more because of retransmissions and acknowledgments.

Multicast Backbone (Mbone), established in 1992, was an experimental network designed to carry IP multicast traffic over the Internet. Multicast can sometimes experience higher latency compared to unicast UDP due to the additional processing required by network devices to manage multicast group memberships and distribute packets to multiple recipients.

For most applications, this doesn't matter. Content Delivery Networks (CDNs) and edge network caching of today's popular content can deliver on-demand content almost near-instantaneously. Think Netflix, etc. And, to be perfectly fair, does it matter if Bob in Dallas gets his on-demand program delivered slower than Sally in Seattle? It's not live, after all.

And let's talk about life. Does Bob know or care that Sally's 'live' program is ahead of his? After all, it's not like they are actually *at* the live event and would notice the difference.

For live events, whether it is a rocket launch, political event, or football game, live streams today are still delivered using packet-based protocols. The packets get reassembled at the user, and it still often takes 30 seconds or even longer for those packets to get put back together and the program to play. If you're not actually there, this doesn't matter.

This lag might not matter when binge-watching a show, but for live events, where fans crave instant engagement, latency is a deal-breaker. Sports betting, for instance, hinges on split-second decisions. A 30-second delay can mean the difference between a winning bet and a useless one. Similarly, fantasy sports players and analysts need real-time player stats as the action unfolds, not 30 seconds later when the game has already moved on.

Of course, if you're *at* the live event, why would you even want to watch or listen to a stream of the event you're physically at? In a lot of cases, that's perfectly true. However, let's say you were at a college or pro football game, and there are eighty thousand people in the stands, and you want to listen to the play-by-play color commentary on the local radio station. Modern stadium architectures often impede AM and FM signal reception.

Beyond commentary, fans expect real-time access to in-depth stats, instant replay angles on demand, and even AI-generated insights piped directly to their devices. The expectation isn't just to watch the game but to experience it with enriched data that keeps up with the on-field action.

Also, who wants to listen to a stream that is thirty seconds or longer out of sync and delayed from the action taking place in front of you? I remember one of the first college games streamed and the wildly enthusiastic reaction from an alumni listening in Japan. He didn't care that the stream had latency – he was just delighted to be able to listen at all!

In recent years, many stadiums have attempted to deploy Wi-Fi infrastructures and partner with tech companies to enable local, real-time streaming: think instant play-by-play commentary or live replays synced with the action. Yet, these efforts often stumble. Network congestion remains a beast: during the 2024 NFL season, some venues saw Wi-Fi grind under 20 TB of data per game, leaving streams buffering at critical moments.

Interference from personal hotspots and IoT gadgets further muddies the signal, pushing latency beyond the 30-second mark fans dread. Add in clunky authentication, like minute-long captive portal logins at places like SoFi Stadium, and spotty coverage in older venues with concrete jungles, and you've got a recipe for fan frustration. Far from seamless, these Wi-Fi woes mean the roar of the crowd often outpaces the stream in your ear.

The impact of these connectivity failures extends beyond just play-by-play audio. Major sportsbooks have invested in live in-game betting, but delays caused by overloaded networks mean odds can shift before a fan's wager even registers. Sports broadcasters and leagues have experimented with interactive elements, such as multiple camera angles or real-time analytics, but without low-latency infrastructure, these features fall flat.

RealNetworks used UDP-based streaming for latency-sensitive services in the 90s, but today's standard on-demand streaming platforms rely primarily on TCP-based HTTP streaming to ensure smooth and reliable playback. UDP remains prominent primarily in real-time, interactive, or latency-critical applications rather than conventional on-demand services.

In the world of streaming, technological advancements have brought us 4K video, real-time gaming, and even immersive AR experiences. Yet, a stubborn issue persists: latency. Live streams often lag 30 seconds or more behind the actual event. For most content, this delay is negligible. But for live sports, betting, and time-sensitive events, it's a critical flaw.

It's not just about listening or watching. It's about engagement. The rise of interactive streaming experiences, such as Twitch-style watch parties for sports, relies on real-time delivery. When a goal is scored, or a touchdown is made, fans expect to react together, not on a 30-second delay. Streaming technology needs to bridge this gap if it hopes to meet the demands of modern sports and entertainment audiences.

This article explores not only the causes of this latency but also examines which platforms use what technologies to address (or not address) this challenge, and why these decisions are made.

How Streaming Platforms Handle Latency

At broadcast.com, as previously discussed, we used both RealNetworks player and server technology as well as Microsoft's. We did play around with Multicast quite a bit. We hoped that it would help solve one of our biggest problems: streaming costs and, secondarily, build a robust distribution infrastructure. In short, however, it had too many problems and never delivered on its promise.

As for lag time, keep in mind this was a time when most people were still on dial-up modems. If you were a geek like Mark or myself, you would have had an ISDN line with an expensive router. Our biggest streams were the occasional 300k, but most were 28.8k and 56k streams. Still, lag time back then was not terribly different than it is now. Just smaller streams that had to be reassembled instead of today's big fat ones. The quality, though, was often poor and nothing remotely like today's '4k' streams.

Different platforms face unique challenges with latency, and their choice of technology often reflects their priorities, whether it's global scalability, low latency, or user experience.

YouTube Live

YouTube Live employs HTTP-based streaming protocols, primarily HTTP Live Streaming (HLS) and Dynamic Adaptive Streaming over HTTP (DASH), to deliver content to a wide array of devices. These protocols are designed for scalability and reliability, ensuring that streams function seamlessly across various platforms. However, their chunk-based architecture can introduce latency ranging from 10 to 30

seconds, which may affect real-time viewer engagement. https://support.
google.com/youtube/answer/7444635

To address latency concerns, YouTube has implemented low latency modes for both HLS and DASH in specific live-streaming scenarios. These enhancements have successfully reduced delays to approximately 5 to 7 seconds, thereby improving the interactivity between streamers and their audiences.

If you're there, though, at a game, 5-7 seconds is a loooong time (much less 30 or more):

- **In five seconds**, an NFL wide receiver can sprint 50 yards downfield, turning a short pass into a dramatic play.
- **In five to seven seconds**, the ball snaps, the quarterback scrambles, evades pressure and launches a deep pass into the end zone.
- **Five seconds** can decide a stolen base, from the pitcher's release, the catcher's throw, to the runner sliding safely under the tag.
- **In seven seconds**, an entire fast break sequence can take place: steal, pass, sprint, and alley-oop dunk.

Who would want to be at the game watching while listening to the play-by-play long after the action is over and the next play has already started?

Despite today's advancements, low-latency streaming is not yet the default setting across the entire platform, indicating ongoing efforts to balance latency reduction with overall stream stability and quality.
https://support.google.com/youtube/answer/7444635

Summary:

Technology Used: HLS (HTTP Live Streaming), DASH (Dynamic Adaptive Streaming over HTTP)

Why: YouTube focuses on scale and reliability. By relying on HTTP-based protocols, YouTube ensures its streams work across a broad range of devices. However, these protocols, with their chunk-based architecture, inherently introduce delays of up to 30 seconds and longer.

Latency Solution Efforts: YouTube has introduced **Low-Latency**

HLS and DASH for certain live-streaming events, reducing delays to as low as 5-7 seconds, but this isn't yet standard across the platform.

Twitch

Twitch employs Low-Latency HTTP Live Streaming (HLS) to facilitate real-time interaction between streamers and viewers, achieving delays as short as 2-3 seconds. This is crucial for maintaining the platform's interactive experience, enabling timely audience engagement during live broadcasts. https://blog.twitch.tv/en/2021/10/25/low-latency-high-reach-creating-an-unparalleled-live-video-streaming-network-at-twitch/

However, prioritizing low latency can lead to increased buffering, particularly for users with unstable or slower internet connections. The reduced buffering window inherent in low-latency streaming allows less time to compensate for network inconsistencies, potentially resulting in playback interruptions. To mitigate this, Twitch offers streamers the option to disable low-latency mode, thereby increasing the buffer duration to enhance playback stability for viewers with suboptimal connections. https://help.twitch.tv/s/article/low-latency-video

Summary:

Technology Used: Low-Latency HLS

Why: Twitch's focus is on real-time engagement between streamers and viewers, making latency a critical issue. By implementing **Low-Latency HLS**, Twitch has reduced delays to as little as 2-3 seconds in many cases.

Challenges: The trade-off is in buffering. Users with poor connections may experience more interruptions as the platform prioritizes low delay over smooth playback.

Zoom

Zoom utilizes Web Real-Time Communication (WebRTC) technology to facilitate its video conferencing services. WebRTC is an open-source framework that enables real-time audio, video, and data sharing directly between browsers and devices without the need for additional plugins.

This technology is designed to support low-latency communication, which is essential for applications like video conferencing, where timely interaction is critical. https://www.wowza.com/blog/what-is-webrtc

However, WebRTC's peer-to-peer architecture presents scalability challenges. While it excels in facilitating direct communication between a limited number of participants, scaling this model to accommodate large audiences, such as in webinars or live broadcasts, can be problematic. The direct connection approach can strain network resources and lead to performance issues as the number of participants increases. https://www.wowza.com/blog/what-is-webrtc

To overcome these limitations, platforms like Zoom implement additional infrastructure, such as Selective Forwarding Units (SFUs), to manage media streams more efficiently. SFUs act as intermediaries that receive media streams from participants and selectively forward them to others, reducing the number of direct connections each participant must manage. This approach helps maintain low latency while improving scalability, allowing platforms to support larger meetings and webinars effectively. https://www. liveswitch.io/blog/how-to-successfully-scale-your-webrtc-application-in-2021

Summary:

Technology Used: WebRTC (Web Real-Time Communication)
Why: For Zoom, low latency is essential. Its reliance on **WebRTC**, a peer-to-peer communication protocol, allows near-instantaneous video and audio delivery for meetings.
Challenges: WebRTC isn't scalable for massive audiences due to its reliance on direct connections, making it unsuitable for platforms like YouTube or Netflix.

Netflix

Netflix employs Dynamic Adaptive Streaming over HTTP (MPEG-DASH) in conjunction with the Transmission Control Protocol (TCP) to deliver its vast library of on-demand content.

MPEG-DASH is an adaptive bitrate streaming technique that enables high-quality streaming of media content over the Internet delivered from conventional HTTP web servers. This combination ensures that each data packet is transmitted accurately, thereby minimizing buffering and optimizing visual quality for viewers. https://en.wikipedia.org/wiki/Dynamic_Adaptive_Streaming_over_HTTP

In the context of on-demand streaming, latency, the delay between a user's action and the response, is less critical compared to live streaming scenarios. A delay of even 30 seconds is inconsequential for Netflix users, as the content is pre-recorded and does not require real-time interaction. Consequently, Netflix prioritizes reliability and high-quality playback over low latency, ensuring a seamless viewing experience for its subscribers.

Summary:

Technology Used: DASH with TCP

Why: Netflix prioritizes reliability and quality over low latency because it specializes in on-demand content. Using **DASH** with TCP ensures every packet of data is delivered accurately, minimizing buffering and optimizing visual quality.

Latency Relevance: For Netflix, a 30-second delay is irrelevant since their content isn't live.

Sports Streaming Platforms (e.g., DAZN, ESPN+)

Sports streaming platforms, such as DAZN and ESPN+, prioritize delivering live content with minimal delay to enhance viewer engagement and maintain the excitement of real-time events. To achieve this, they employ technologies like Low-Latency HTTP Live Streaming (LL-HLS), Dynamic Adaptive Streaming over HTTP (DASH), and proprietary protocols designed to reduce latency. For instance, DAZN has expanded its partnership with LTN to deliver low-latency live streaming channels to betting locations, utilizing LTN's managed IP network that offers 99.999% reliability and sub-300ms latency. https://www.sportsvideo.org/2024/12/10/

dazn-ltn-expand-partnership-to-deliver-11-low-latency-live-streaming-channels-to-betting-locations/

Despite these technological advancements, maintaining consistent low-latency streams for millions of simultaneous viewers presents significant technical and economic challenges. Scaling infrastructure to handle high traffic volumes without compromising stream quality requires substantial investment in robust content delivery networks (CDNs) and efficient encoding processes. Additionally, varying network conditions across different regions can lead to buffering or reduced video quality, impacting the overall viewer experience. https://www.sportspro.com/insights/gcore-cdn-tech-streaming-ultra-low-latency/

Summary:

Technology Used: Low-Latency HLS, DASH, Proprietary Protocols

Why: Sports streaming platforms face the dual challenge of maintaining high-quality streams while minimizing delay. Many use **Low-Latency HLS** or proprietary technologies designed to reduce chunk sizes and optimize CDN delivery.

Challenges: Achieving consistency for millions of viewers watching simultaneously remains a technical and economic hurdle.

Real-Time Betting Platforms

Real-time betting platforms prioritize ultra-low latency to ensure fair play and maintain the integrity of live wagering. To achieve this, they commonly utilize the Real-Time Transport Protocol (RTP) and Web Real-Time Communication (WebRTC). RTP facilitates the rapid transmission of audio and video data, while WebRTC enables peer-to-peer communication directly between browsers, minimizing delay. These technologies are essential for delivering the immediacy required in live betting scenarios. https://www.wowza.com/blog/streaming-protocols

However, scaling these technologies to accommodate large audiences presents significant challenges. WebRTC, for instance, was originally designed for direct communication between a limited number of participants, and scaling it for mass streaming requires complex infrastructure and substantial computational resources. This complexity can lead to increased costs and potential performance issues, making it less suitable for mainstream streaming platforms that serve vast audiences. https://www.streamingmedia.com/Articles/ReadArticle.aspx

To overcome these limitations, some platforms are exploring hybrid approaches that combine WebRTC with other streaming technologies or implementing advanced architectures to enhance scalability. Despite these efforts, achieving the necessary scale for widespread adoption in mainstream streaming remains a complex and resource-intensive endeavor....https://blog.phenixrts.com/a-better-way-to-scale-webrtc

Summary:

Technology Used: RTP (Real-Time Transport Protocol), WebRTC

Why: Platforms that support live betting prioritize ultra-low latency to ensure fair play. RTP and WebRTC are commonly used due to their ability to transmit data with minimal delay.

Challenges: These technologies struggle to scale beyond small, dedicated audiences, making them unsuitable for mainstream streaming.

Why Platforms Make These Choices

Streaming platforms carefully select their streaming technologies based on specific priorities and operational considerations:

Scalability vs. Real-Time Needs

Platforms like YouTube and Netflix prioritize scalability and reliability to serve a global audience. Their primary focus is on delivering high-quality content efficiently, where real-time interaction is not essential. Consequently, they utilize protocols such as HTTP Live Streaming (HLS) and Dynamic Adaptive Streaming over HTTP (DASH), which, while introducing higher latency, offer robust performance and broad compatibility across devices. https://www.wowza.com/blog/streaming-protocols-latency

Interactivity and Community Engagement

Platforms such as Twitch, Zoom, and real-time betting services cater to users who require immediate feedback and real-time interaction. To facilitate this, they employ low-latency streaming protocols like Web Real-Time Communication (WebRTC) and Real-Time Transport Protocol (RTP). These protocols minimize delay, enhancing user engagement and interactivity. However, achieving low latency often involves trade-offs in terms of reliability and compatibility, as these protocols may be more susceptible to network variations and may not be supported across all devices....https://www.gumlet.com/learn/webrtc-vs-hls/

Infrastructure Costs

Implementing low-latency protocols such as WebRTC and RTP necessitates substantial investment in infrastructure to ensure scalability and performance. The peer-to-peer architecture of WebRTC, for example, can lead to increased complexity and resource consumption when scaling to large audiences. For many platforms, especially those serving vast user bases, the costs associated with deploying and maintaining such technologies on a global scale may outweigh the benefits of reduced latency. As a result, they opt for more scalable solutions that, while introducing higher latency, offer greater reliability and cost-effectiveness. https://cloudinary.com/guides/live-streaming-video/low-latency-hls-ll-hls-cmaf-and-webrtc-which-is-best

Emerging Solutions:

Who's Leading the Way?

To address the challenge of balancing scalability with real-time performance, streaming platforms are investing in innovative technologies aimed at reducing latency and enhancing user experiences.

Edge Computing

Platforms such as Twitch and ESPN+ are adopting edge computing to process data closer to viewers, thereby reducing the physical distance content must travel. By decentralizing data processing and bringing it to the network's edge, these platforms can minimize latency introduced by traditional Content Delivery Network (CDN) routing. This approach not only reduces latency but also improves bandwidth efficiency and overall performance, offering a pathway toward real-time streaming at scale. https://www.muvi.com/blogs/role-of-edge-computing-in-video-streaming/

AI-Driven Optimization

Streaming services are increasingly leveraging artificial intelligence (AI) to optimize various aspects of content delivery. For instance, platforms like YouTube utilize AI to adjust streaming quality in real-time based on the viewer's internet connection, ensuring smooth playback even on low-bandwidth connections. Additionally, AI is being explored to predict and mitigate network congestion, helping to deliver low-latency streams without compromising quality. https://www.forbes.com/sites/neilsahota/2024/03/18/streaming-into-the-future-how-ai-is-reshaping-entertainment/

Proprietary Protocols

Some sports streaming platforms, such as DAZN, are experimenting with proprietary protocols that combine elements of existing streaming technologies to balance low latency with reliability. For example, DAZN has expanded its partnership with LTN to deliver low-latency live streaming channels to betting locations, utilizing LTN's managed IP network that offers sub-300ms latency. While promising, these solutions are often confined within specific ecosystems, limiting broader adoption. https://www.forbes.com/sites/ neilsahota/2024/03/18/streaming-into-the-future-how-ai-is-reshaping-entertainment/

5G Networks

The advent of 5G technology presents significant opportunities for mobile streaming platforms. With its ultra-low latency capabilities, 5G enables real-time streaming experiences that were previously unattainable on mobile devices. Platforms like Twitch and YouTube Live are piloting 5G-powered real-time streams in select regions, aiming to enhance mobile-first applications such as augmented reality (AR) and live sports betting. This development could be a game-changer, providing the necessary infrastructure to support high-quality, low-latency streaming on the go. https://www.verizon.com/business/resources/articles/s/ how-edge-computing-can-help-improve-audio-and-video-streaming-technology/

In summary, by investing in edge computing, AI-driven optimization, proprietary protocols, and 5G networks, streaming platforms are taking bold steps to bridge the gap between scalability and real-time performance, ultimately enhancing the viewing experience for their audiences.

Conclusion: The Latency Balancing Act

Latency remains one of the most persistent challenges in the streaming industry. While some platforms have made strides toward reducing delays, the trade-offs between scalability, reliability, and cost continue to hinder the widespread adoption of ultra-low-latency solutions.

As new technologies like edge computing, AI-driven optimization, and 5G gain traction, the gap is closing, but at a slow pace. Solving the latency problem at scale will require collaboration across industries and significant investment in infrastructure. Until then, platforms will continue to navigate the balancing act between what's possible and what's practical.

What do you think will drive the next big leap in low-latency streaming?

12: From Broadcast.com to ESPN+

The Business of Direct-to-Consumer Sports

The business of live sports broadcasting is at a crossroads. Gone are the days when traditional cable packages and predictable revenue streams defined the industry. Today, the convergence of skyrocketing media rights fees, rapidly evolving consumer preferences, and the disruptive power of streaming technology have set broadcasters and rights holders on a challenging yet exciting new path.

This chapter delves deeply into the shifting landscape of live sports economics, analyzing how escalating rights fees and shifting viewer behaviors are reshaping the industry's financial fundamentals. We'll examine how the resurgence of pay-per-view (PPV) models is both a lifeline and a complication for broadcasters, reflecting both the value of high-profile, real-time events and the growing consumer frustration over costs.

Additionally, we explore the rising prominence of direct-to-consumer (DTC) platforms like ESPN+, DAZN, and league-specific streaming services, which are redefining relationships between leagues, broadcasters, and fans. These platforms offer unprecedented opportunities for personalization and data-driven engagement but also present significant challenges, from managing scalability during peak demand to navigating complex consumer relationships.

From the pioneering days of Broadcast.com's streaming of the Super Bowls in the late 1990s, complete with chaotic server rooms and a pioneering spirit, to today's billion-dollar media rights deals and global streaming networks, the evolution has been dramatic and instructive. This chapter will take you behind the scenes, shedding light on the successes, struggles, and strategic pivots that define the contemporary sports media landscape.

The Skyrocketing Cost of Sports Rights

At the heart of the sports streaming industry's challenges lies the astronomical cost of securing broadcast rights. These rights are the lifeblood of sports broadcasters, but their price tags have reached unprecedented levels.

The Billion-Dollar Leagues

The escalating costs of securing sports broadcasting rights have become a defining challenge for the industry, with major leagues commanding unprecedented sums for media access.

National Football League (NFL):

Broadcast.com carried Super Bowls XXX (January 28, 1996), XXXI (January 26, 1997), XXXII (January 25, 1998), and XXXIII (January 31, 1999). The first of these, Super Bowl XXX (Pittsburgh Steelers vs. Dallas Cowboys), was AudioNet's initial Super Bowl broadcast. Back then, saying that events like the Super Bowl required all hands on deck was the understatement of the decade. Our broadcast/data center was a madhouse of criss-crossed cables, repatched and repurposed hardware, and fevered rebalancing of our ever-expanding network of edge-located servers around the globe.

Part of our business model was signing up radio stations that had the audio rights to various games from various leagues. Back then, nobody cared about those rights. So, we offered to put the radio station up on the Internet in exchange for some of their monthly unsold spot inventory – and the ability to carry the audio of the games they carried.

Eventually, our rapid growth put us squarely on the NFL's radar, and predictably, they weren't thrilled. We soon received the inevitable cease-and-desist order. Without missing a beat, Mark Cuban flew straight to New York to meet with them face-to-face. Not only did he defuse a potentially serious confrontation, but he also somehow managed to completely turn the situation around. Say what you will about Mark, but I've always admired his remarkable talent for negotiation – his uncanny ability to step into challenging situations, transform adversaries into allies, and create opportunities where others saw only roadblocks. This incident with the NFL was a perfect example of how effectively Mark could navigate tense negotiations and emerge not only unscathed but with new partners in tow.

Let me be clear: I'm no Mark Cuban fanboy. I spent enough time working closely with him to develop a far more nuanced opinion. Mark was undeniably a natural-born salesman, brimming with energy and confidence, always eager to make bold promises and sweeping commitments. However, for those of us behind the scenes, his charismatic assurances often felt like ticking time bombs. Frequently, we'd find ourselves scrambling frantically, putting out fires and racing to deliver on promises that, truthfully, we weren't entirely sure we could keep. Navigating those situations was both exhausting and exhilarating – always challenging, sometimes chaotic, but never dull.

Since those pioneering days of streaming experimentation, the NFL's broadcasting landscape has transformed dramatically. In 1990, the league ratified four-year television agreements with ABC, CBS, NBC, ESPN, and TNT totaling $3.6 billion, marking the largest sports media package in television history at the time. By contrast, in 2021, the NFL finalized an 11-year media rights agreement valued at approximately $110 billion, averaging over $10 billion annually. This exponential growth underscores the NFL's rising prominence and the escalating value placed on live sports content....https://en.wikipedia.org/wiki/NFL_on_American_television, https://en.wikipedia.org/wiki/History_of_the_NFL_on_television

National Basketball Association (NBA):

The NBA has secured an 11-year media rights deal worth $77 billion with ESPN, NBCUniversal, and Amazon, set to commence in the 2025-26 season. This agreement reflects the league's growing popularity and the increasing value of live sports content.
https://www.sportsbusinessjournal.com/Articles/2024/07/10/nba-media-rights-streaming-how-to-watch

FIFA World Cup:

Broadcast rights for the FIFA World Cup command substantial investments every four years. For instance, Fox and Telemundo collectively paid $1 billion for the rights to the 2026 tournament, underscoring the event's global appeal and the premium placed on exclusive broadcasting access.

As a side note, I vividly remember my trip to Paris for the World Cup in 1998. At the time, both the government of France and the City of Paris had extended invitations for me to visit and explore the possibility of opening a Broadcast.com office there. Among the unforgettable highlights was attending the World Cup match, where France secured a historic victory. The entire city erupted in joyous chaos – an absolute frenzy of celebration unlike anything I'd ever experienced.

Even my VIP bus wasn't immune; it was enthusiastically rocked side-to-side by ecstatic fans and nearly overturned. With no other choice, we abandoned the bus and set off on foot, making our way through streets packed shoulder-to-shoulder with jubilant Parisians. Navigating through that sea of euphoria was surreal, exhilarating, and unforgettable – truly an extraordinary night.

Interestingly, during that same memorable visit, we were invited to an exclusive wine cellar for a private tasting event. Imagine my astonishment – and delight – when I discovered that the talented sommelier guiding us through the selections was not only incredibly knowledgeable but also a charming, sophisticated, and strikingly beautiful dark-skinned woman originally from Houston. It was one of those remarkable, unexpected encounters that somehow made Paris feel momentarily familiar, turning an elegant evening into an even more unforgettable experience.

Major League Baseball (MLB):

Regional Sports Networks (RSNs) like Bally Sports have heavily invested in MLB broadcasting rights. However, the combination of rising rights costs and declining viewership has placed financial strain on these networks, leading to concerns about the sustainability of such investments.

These escalating costs reflect the intense competition among broadcasters and streaming platforms to secure exclusive sports content, which remains a key driver of subscriptions and advertising revenue. However, the substantial financial commitments required pose significant challenges, particularly as consumer viewing habits evolve and the media landscape continues to shift.

Why Costs Keep Rising

- **Scarcity of Live Sports:** Unlike scripted entertainment, sports are unscripted, real-time events that demand live viewership, making them a premium commodity in the age of on-demand content.
- **Global Demand:** Sports leagues are expanding their reach to global audiences, further driving up the value of their rights.
- **Bidding Wars:** The competition between traditional broadcasters (CBS, NBC, ESPN) and streaming platforms (Amazon Prime, Apple TV+, YouTube) has intensified, inflating prices.

The Return of Pay-Per-View

The escalating costs of sports broadcasting rights have prompted a resurgence in the pay-per-view (PPV) model as broadcasters and streaming platforms seek viable methods to recoup their substantial investments.

Why PPV is Making a Comeback

High-Profile Events

Boxing and Mixed Martial Arts (MMA): PPV remains a dominant revenue model for marquee boxing and MMA events. For instance, the anticipated bout between Mike Tyson and Jake Paul streamed on Netflix, attracted significant attention, with both fighters reportedly earning over $20 million each. https://www.theguardian.com/sport/2024/nov/10/mike-tyson-jake-paul-boxing-netflix

Selective Live Sports: Streaming platforms are exploring PPV for exclusive live sports content. Netflix's venture into live sports, exemplified by the Tyson-Paul fight, indicates a strategic move to offer premium events on a PPV basis, providing subscribers with unique, high-profile content. https://www.investopedia.com/netflix-likes-streaming-live-events-jake-paul-mike-tyson-nfl-8778375

Supplemental Revenue Streams

For platforms that have invested heavily in sports rights, PPV offers an avenue to generate additional revenue without imposing higher subscription fees across their entire user base. This model allows for the monetization of premium events, catering to viewers willing to pay extra for exclusive content. https://blog.onstreammedia.com/ppv-live-streaming-services

Challenges with PPV

Consumer Pushback: The frequent implementation of PPV can lead to consumer dissatisfaction, especially among viewers already managing multiple subscription services. The additional cost of PPV events may be perceived as burdensome, potentially leading to subscriber attrition. https://www.boxingscene.com/pay-per-view-piracy-reaching-crisis-point--184223

Piracy Concerns: High PPV prices often drive consumers toward illegal streaming alternatives. Studies indicate that approximately 11% of U.S. adults admit to pirating live sports, TV, or movies, citing high costs as the primary motivation. This trend not only undermines revenue but also poses legal and ethical challenges for broadcasters. https://ipwatchdog.com/2025/01/23/ip-octagon-saving-live-sports-digital-piracy/id%3D185316/

In summary, while the PPV model provides a mechanism for broadcasters and platforms to offset the rising costs of sports rights, it must be balanced carefully against potential consumer resistance and the pervasive issue of digital piracy. Strategic pricing and value propositions are essential to ensure that PPV offerings remain attractive and accessible to a broad audience.

From Cable to OTT: Tales from the Trenches

The call came in 2014: another "emergency" project. You know the type: "90 days to launch," "everything's broken," and "we need help yesterday." This time, a Mexican cable operator needed to launch their OTT (Over The Top) streaming service. The scope? Converting 50 linear TV channels to live streams, plus video-on-demand capabilities, all accessible on mobile devices and online.

That's a combination of online platforms as well as physical encoding / transcoding equipment and servers onsite at the broadcast center in central Mexico.

In time for the World Cup.
In 90 Days.
Deep Breath....
I could tell you how we got it done, but I'd have
to.... Well, that's a story for another day.

As many of you know, 90-day crash projects are often not just ambitious. They are often very nearly impossible. But "impossible" projects had become something of a specialty of mine since the AudioNet days.

Besides the mad deadlines needed to build the online infrastructure, there was the minor challenge of getting the necessary equipment into Mexico. Getting it, there may or may not have involved certain extremely... creative... methods of getting the computers and broadcast equipment across the border in time to build things out under... austere... conditions.

Anyone who's set up broadcasting equipment in challenging environments knows that every day brings a new crisis to solve. But the team was incredible, and somehow, through long days, longer nights, and enough coffee to float a battleship, we pulled it off.

We launched just in time for *World Cup* kickoff. I've never seen beer disappear so fast at a launch party, and for good reason. The platform worked flawlessly.

The success in Mexico led to another opportunity: bringing similar capabilities to the Philippines. A cable operator there needed their own OTT and VOD platform, again with 50 linear channels. While the conditions weren't quite as "interesting" as Mexico, spending several months at a former airbase had its own unique challenges.

The Philippine project was particularly memorable for the people. Their technical teams were eager to learn, genuinely hospitable, and brought an energy to the project that made those long deployment days feel shorter. Plus, their local hospitality and food beat typical data center vending machine fare by several orders of magnitude.

These projects reinforced something I'd learned at Broadcast.com: in streaming, the impossible just takes a little longer and sometimes requires creative problem-solving that's best not discussed in detail.

The Shift to Direct-to-Consumer (DTC) Streaming Models

The sports industry is experiencing a significant transition from traditional linear television to direct-to-consumer (DTC) streaming platforms, enabling leagues and broadcasters to manage their content and audience relationships more directly.

Key Players in the DTC Revolution

ESPN+: ESPN's streaming service offers a blend of live sports, exclusive events, and original programming, attracting a substantial subscriber base. However, it faces the ongoing challenge of balancing affordability for consumers with the high costs associated with securing broadcasting rights.

NFL+: Launched in 2022, NFL+ provides fans with access to live local and primetime games on mobile devices, granting the league direct control over its digital audience. This move reflects the NFL's strategy to engage viewers more intimately and adapt to changing consumption habits.

DAZN: Often referred to as the "Netflix of Sports," DAZN specializes in DTC streaming, offering subscriptions for sports such as boxing and soccer. Despite significant investments, profitability remains a challenge due to the high costs of rights deals. Notably, Sir Leonard Blavatnik has invested an additional $827 million into DAZN, bringing his total investment to over $6.7 billion since its inception in 2016. DAZN reported revenue growth of $2.9 billion in 2023, with projections to exceed $6 billion by 2025, partly due to strategic acquisitions and broadcasting rights deals. https://www.ft.com/content/4aff9714-f70f-4ce1-ac8d-f2fcf682a434

This shift towards DTC models reflects a broader trend in the sports industry, as organizations seek to leverage digital platforms to enhance viewer engagement, gather audience data, and create personalized experiences. However, the transition also presents challenges, including the need to manage substantial rights costs, technological infrastructure, and competition from traditional broadcasters and emerging platforms.

As the landscape continues to evolve, the success of DTC streaming services will depend on their ability to offer compelling content, maintain cost-effective operations, and adapt to the dynamic preferences of sports audiences.

Why the DTC Shift Makes Sense

The transition to direct-to-consumer (DTC) streaming models in the sports industry offers several compelling advantages:

1. Audience Engagement
DTC platforms enable sports leagues and organizations to establish direct relationships with fans, allowing for the collection of valuable data to personalize experiences and foster loyalty. By understanding viewer preferences and behaviors, these platforms can tailor content, promotions, and interactions to enhance fan engagement. https://www.pwc.com/us/en/industries/tmt/library/sports-streaming-platforms.html

2. Revenue Diversification
Implementing a DTC model opens multiple revenue streams beyond traditional broadcasting. These include subscription fees, pay-per-view events, in-platform advertising, and merchandise sales. By diversifying income sources, sports entities can reduce reliance on singular revenue channels and create more sustainable financial models. https://www.sportsbusinessjournal.com/Articles/2024/07/16/oped-16-salah-morgan

3. Global Reach
Streaming platforms have the capability to bypass regional restrictions, allowing sports content to reach a worldwide audience without the need for intermediaries. This global accessibility not only expands the fan base but also increases potential revenue from international markets. https://about.grabyo.com/future-of-sports-streaming/

By embracing DTC streaming models, sports organizations can enhance fan engagement, diversify revenue streams, and extend their reach to a global audience, aligning with the evolving consumption habits of modern viewers.

The Challenges of Going Direct

Transitioning to direct-to-consumer (DTC) streaming models presents several challenges for sports leagues and broadcasters:

1. Fragmentation
As more leagues and organizations launch their own DTC

platforms, fans are required to manage multiple subscriptions to access all their desired content. This proliferation of services can lead to consumer frustration, as viewers must navigate various platforms and incur additional costs to follow their favorite sports. A study highlighted that 69% of avid sports fans find using multiple providers to watch the same sport burdensome, and 59% report increased difficulty in locating the sports content they wish to watch. https://www.newscaststudio.com/2024/08/28/sports-streaming-gains-but-viewers-struggle-with-fragmentation/

2. Scalability

Ensuring that streaming infrastructure can handle peak traffic during high-profile events is a significant concern. Live sports streaming demands robust infrastructure capable of delivering content seamlessly to millions of concurrent viewers. Challenges such as device and network compatibility, latency issues, and maintaining reliability under high demand necessitate continuous testing and optimization. For instance, during major tournaments like March Madness, streaming platforms are put to the test to manage multiple games across different networks and platforms simultaneously. https://www.streamingmedia.com/Articles/Post/Blog/The-Complexities-of-Sports-Streaming-Requires-Continuous-Testing--164428.aspx

Addressing these challenges is crucial for the successful implementation of DTC streaming strategies in the sports industry. Organizations must focus on enhancing user experience by simplifying access and investing in scalable, resilient infrastructure to meet the demands of live sports streaming.

Conclusion: Navigating the Future of Sports Streaming

The evolution of sports broadcasting from those chaotic server rooms at Broadcast.com to today's sophisticated streaming empires represents more than just technological advancement. It's a fundamental reimagining of how fans consume sports and how the industry monetizes that passion.

We've come a long way from patching cables and repurposing hardware to stream Super Bowl XXX. Back then, we were flying under the radar, converting radio rights nobody cared about into streaming gold. Today, those same rights command billions, with the NFL alone securing $110 billion over 11 years, a staggering leap from the $3.6 billion deals of the early 1990s.

This explosion in rights fees has triggered a complex chain reaction. Traditional broadcasters are scrambling to justify these investments while streaming natives like DAZN burn through billions chasing profitability. Meanwhile, leagues themselves are bypassing middlemen entirely with services like NFL+ and NBA League Pass, creating direct relationships with fans that were unimaginable in the AudioNet days.

The resurgence of pay-per-view models for premium events represents both opportunity and threat. While it provides a revenue lifeline, each additional charge risks pushing fans toward piracy, that persistent challenge that's followed us from the days of password sharing to today's sophisticated illegal streams.

Perhaps the most fascinating evolution is how leagues and broadcasters are navigating the scalability challenges that once had us sweating in server rooms. Today's "90-day crash projects" involve complex global infrastructure that must reliably deliver 4K streams to millions of concurrent viewers, whether they're watching World Cup matches in Mexico or NBA games in the Philippines.

The fragmentation problem looms large, too. When fans need five different subscriptions to follow their favorite sports, we're testing their patience and wallets. Finding the sweet spot between profitability and accessibility remains the industry's great unsolved puzzle.

As someone who's lived through this transformation from the wild-west days of internet broadcasting to today's sophisticated streaming landscape, I can tell you this: the challenges have grown more complex, but the fundamentals haven't changed. Success still requires creative problem-solving, sometimes with methods "best not discussed in detail," and the willingness to tackle the seemingly impossible with coffee, determination, and a bit of madness.

The sports streaming industry stands at a fascinating inflection point. Those who can balance astronomical rights fees with compelling consumer offerings and who can build infrastructure robust enough for Super Bowl-scale events while maintaining profitability will define the next chapter. The scrappy spirit that drove us to string cables across makeshift server rooms decades ago is still the same spirit driving innovation today, just with considerably higher stakes and far better technology.

What do you think the sports streaming industry's next big innovation will be?

13: Beyond Broadcast.com

How Influencers and Algorithms Sparked an Audience Rebellion

Not that long ago, in a galaxy not very far away, TV executives and Hollywood gatekeepers had complete and utter control over what and when we watched. They alone wielded the power to decide who became a star and what films and shows made it to our screens.

Then came the '90s – a New Hope. The rise of the internet, championed by rebels like AudioNet.com (later Broadcast.com), challenged the empire's rule, granting viewers the revolutionary freedom to watch or listen to whatever they wanted, whenever and wherever they chose.

Introduction: The Rise of the Algorithm as Gatekeeper

As broadband spread faster and stronger than ever, a powerful new Force emerged – streaming platforms such as Netflix, YouTube, and

TikTok, altering the media universe forever. But it wasn't long before these platforms, tempted by the Dark Side, harnessed the power of algorithmic control. Quietly, they began guiding which content would soar through hyperspace to success and which would vanish into the void.

For years, the ~~Dark Side~~ tech platforms operated almost unchecked, gathering data, analyzing our viewing habits, and tweaking their recommendations to maximize our screen time. They had the power.

For years, these tech Empires dominated with seemingly unstoppable control, gathering data, meticulously studying our viewing habits, perhaps even listening to us on our phones and devices, and fine-tuning their recommendations to maximize viewer engagement. Yet, in true rebellion fashion, a new disturbance emerged in the Force – the audience itself.

Viewers have uncovered ways to disrupt the algorithmic Empire's hold, turning platforms' own tools against them through tactics like "ratio-ing," meme warfare, and influencer-driven rebellions. Algorithms designed to control users have found themselves at the mercy of these very same rebels.

Now, the balance of power teeters precariously once again. Is this a new Jedi era, where the Force – viewer choice – finally reigns supreme? Or have we simply replaced one Imperial regime with another, where the crowd has become the collective Emperor?

What does this mean for the future of streaming, film, and even live sports?

In this chapter, we'll journey deep into how algorithms, user rebellion, and viral social media uprisings are reshaping the galactic distribution of content. We'll examine the hidden Death Star-like strategies streaming platforms employ, how audiences influence content destiny, and how even live sports broadcasting faces its own Galactic Civil War in the eternal struggle for viewer attention.

Algorithmic Control: How Platforms Decide What We See

The algorithm is the invisible hand that guides our digital consumption. Whether we're watching a show on Netflix, browsing videos on YouTube, or scrolling through TikTok, an AI-driven system is making choices for us. It analyzes watch time, engagement, and historical behavior to serve up content that will keep us hooked. But while this system is designed to maximize engagement, it is also a tool that platforms use to shape the content landscape in their favor.

Netflix's Data-Driven Content Prioritization

Netflix doesn't just recommend content; it attempts to engineer success. The platform's algorithm is notorious for promoting its own productions over licensed content, ensuring that original programming like *Stranger Things* or *Squid Game* gets front-page exposure. Shows that underperform in their first few days or weeks are often quietly buried, never receiving another push in the recommendation engine, and their renewals canceled. In this way, Netflix has turned data into a predictive model for content success, attempting to decide which series will thrive and which will be forgotten.

YouTube's Engagement-Based Algorithm

YouTube's recommendation engine operates on a different model, one that prioritizes watch time over everything else. The more time users spend watching a particular type of content, the more the algorithm pushes similar videos. This system has led to the rise of hyper-optimized content designed to exploit engagement metrics.

Controversial content, extreme opinions, and sensational headlines, when not outright censored by the algorithm, often outperform nuanced discussions because they generate more clicks, comments, and shares. Additionally, YouTube is somewhat notorious for attempting to shape narratives by demonetizing content based on opaque internal, sometimes sociopolitical rules driven by parent company Google's imperatives.

TikTok's Viral Amplification

TikTok's For You Page (FYP) is one of the most powerful recommendation engines in the world. Unlike Netflix or YouTube, TikTok doesn't rely on user subscriptions or extensive watch history. Instead, it monitors immediate engagement, likes, shares, and watch duration to determine how widely a video should be distributed. This rapid feedback loop has made TikTok the birthplace of viral trends, but it has also created a platform where fame can be fleeting, and content can be buried just as quickly as it rises.

Search Manipulation & Shadow Suppression

Beyond recommendations, platforms also manipulate what users find in search results. Google, YouTube, and social media platforms have all been accused of shadow suppression, deliberately pushing certain content down in rankings or removing it entirely from search visibility. Whether done for political reasons, brand protection, or content moderation, this practice raises the question: *Are we really choosing what we watch, or are the platforms choosing for us?*

As streaming platforms and social media giants continue to refine their algorithms, they are also facing new challenges from an audience that is learning how to fight back. The next section will explore how users are hijacking engagement metrics through ratio-ing, influencer manipulation, and audience-driven discourse to shape content success or failure.

The Rise of Audience Hijacking: Ratio-ing, Downvotes, and Podcasters

Ratio-ing: The Audience's New Weapon

Ratio-ing has become a powerful tool for audiences to collectively suppress or elevate content. Originally a Twitter (X) phenomenon where a post received significantly more comments than likes, indicating backlash, the concept has now expanded to YouTube, TikTok, and other platforms.

On YouTube, a high dislike-to-like ratio can signal mass disapproval, leading the algorithm to suppress visibility. Movie trailers, political speeches, and corporate announcements have all been targets of ratioing, where overwhelming negative engagement drowns out the original content's intended message. One approach to combat these efforts is to simply disable up/down votes and comments....

Case Study: Disney's Snow White (2024-2025)

In 2024 and continuing into 2025, Disney's live-action _Snow White_ remake became a focal point of online controversy, with its early trailers receiving significant backlash on social media platforms. The criticism centered on perceived deviations from the original narrative and character portrayals, which many felt were driven by contemporary political agendas. This sentiment was exacerbated by statements from the film's lead actress, Rachel Zegler, that went viral and were met with widespread disapproval. The backlash was further amplified by podcasters, YouTubers, and social media influencers, leading to extensive online debates and discussions.

The controversy primarily revolved around significant departures from the original story, including the reimagining of key characters in ways that many perceived as politically motivated. Critics accused Disney of prioritizing modern messaging over a faithful adaptation, sparking widespread online debate.

In response to the mounting criticism, Disney delayed the film's release by a year, reportedly to address the feedback and make necessary changes. The production underwent significant reshoots and alterations to better align with audience expectations.

This incident underscores a significant shift in the entertainment industry: studios and streaming platforms can no longer rely solely on traditional marketing strategies. They must adeptly navigate the unpredictable realm of user engagement, where audience sentiment, amplified through digital platforms, can profoundly influence a project's trajectory even before its official release. https://www.vanityfair.com/hollywood/story/the-curse-of-snow-white-all-the-drama-around-disneys-remake-explained, https://www.thesun.co.uk/tv/33844272/disney-woke-snow-white-dwarfs-feud/ , https://nypost.com/2025/03/16/entertainment/rachel-zegler-gal-gadot-attend-snow-white-premiere-in-los-angeles/

Case Study: Marvel's Secret Invasion Opening Credits (2023)

In 2023, Marvel Studios' series *Secret Invasion* faced significant backlash over its AI-generated opening credits. This decision drew criticism from fans and industry professionals alike, particularly as it coincided with ongoing Hollywood labor disputes, including the Writers Guild of America (WGA) strike, where the role of AI in creative fields was a central issue. https://ew.com/tv/secret-invasion-marvel-ai-generated-intro-controversy/

The opening sequence, crafted by Method Studios using generative AI, was intended to reflect the show's themes of shape-shifting and identity. Director Ali Selim explained that the AI-generated visuals emerged from the "shape-shifting, Skrull world identity" central to the series' narrative. https://www.polygon.com/23767640/ai-mcu-secret-invasion-opening-credits, https://ew.com/tv/secret-invasion-marvel-ai-generated-intro-controversy/

However, this artistic choice was met with widespread disapproval. Many viewers and artists expressed concerns that employing AI for the credits undermined human creativity and potentially threatened jobs in the creative industry. Storyboard artist Jon Lam described the move as "a slap in the face of artists and the WGA strike," highlighting the ethical implications of replacing human labor with AI. https://news.artnet.com/art-world/marvel-secret-invasion-ai-opening-credits-2325932 , https://timesofindia.indiatimes.com/gadgets-news/marvel-studios-secret-invasion-faces-backlash-over-ai-generated-scenes-heres-what-the-creators-has-to-say/articleshow/101222684.cms , https://www.cartoonbrew.com/series/marvel-secret-invasion-ai-opening-credits-229988.html

In response to the criticism, Method Studios clarified that the AI tools were just one component of the creative process and did not replace any artists' jobs. They emphasized that the production was "highly collaborative and iterative," involving a team of art directors, animators, and developers who used both conventional techniques and AI tools to achieve the desired aesthetic. https://www.ign.com/articles/studio-behind-secret-invasions-ai-intro-addresses-backlash-no-artists-jobs-were-replaced, https://www.polygon.com/23767640/ai-mcu-secret-invasion-opening-credits

This incident underscores the complex debate surrounding AI's role in creative industries. While some view AI as a tool that can augment artistic expression, others see it as a potential threat to employment and the authenticity of human creativity. The timing of Marvel's decision, amid labor disputes focusing on AI's impact on jobs, further intensified the controversy, leading to discussions about ethical production choices in modern media.

This case exemplifies how corporate decisions to incorporate AI can backfire, especially when audiences are increasingly vigilant about the implications of technology on employment and creativity. It highlights the necessity for studios to consider public sentiment and ethical considerations when integrating new technologies into creative processes.

Case Study: Indiana Jones and the Dial of Destiny (2023) – Too Big to Fail?

Production Budget and Box Office Performance

Indiana Jones and the Dial of Destiny had an estimated production budget of $294.7 million, positioning it among the most expensive films ever produced. Despite the franchise's legacy and Disney's high expectations, the film grossed approximately $384 million worldwide, falling short of its estimated $600 million break-even point. https://www.boxofficemojo.com/franchise/fr3209137925/

Critical Reception and Fan Backlash

The film faced criticism for its narrative choices, pacing, and the diminished role of Harrison Ford's iconic character. These creative decisions led to disappointment among long-time fans, who felt alienated by the attempt to modernize the beloved archaeologist. Director James Mangold expressed his hurt over the negative reception, acknowledging the challenges in meeting audience expectations. https://collider.com/indiana-jones-and-the-dial-of-destiny-james-mangold-reacts-audience-reception/

Pre-Release Skepticism and Digital Discourse

Prior to its release, the announcement of a fifth installment sparked skepticism across social media platforms. Discussions questioned the necessity of continuing the series, with concerns about potential deviations from the original spirit of the franchise. This pre-release negativity contributed to lukewarm audience reception and underwhelming box office results.

Financial Implications and Industry Lessons

The film's performance resulted in significant financial losses for Disney, estimated at around $100 million. This outcome challenges the assumption that substantial budgets and established franchises inherently guarantee success. The case underscores the importance of aligning creative decisions with fan expectations and engaging authentically with digital communities to foster positive reception. https://screenrant.com/ indiana-jones-5-movie-box-office-profit-loss/

This scenario illustrates the "Too Big to Fail" fallacy, highlighting that financial investment and brand recognition alone cannot ensure a film's success. Studios must consider fan sentiment, community engagement, and digital discourse as critical factors influencing a project's outcome.

Key Takeaways from These Case Studies

These examples highlight three major trends in modern media reception:

1. **Ratio-ing and Review Bombing as Audience-Led Resistance**
   - Social media platforms allow communities to collectively push back against films and TV series in ways that studios cannot easily counter.
   - The backlash against *Snow White* and *Emilia Pérez* demonstrates how pre-release narratives can doom a project before it even reaches audiences.
2. **The Rise of Ethical and Political Controversies in Media**
   - *Secret Invasion* and *Emilia Pérez* show that creative and business decisions can become lightning rods for public debate.
   - Whether it's AI replacing human animators or concerns over cultural and gender representation, studios now face a hyper-aware and hyper-critical audience.
3. **The "Too Big to Fail" Fallacy in High-Budget Productions**
   - *Indiana Jones and the Dial of Destiny* and *The Rings of Power* both suffered from audience skepticism despite massive investments.

o   These cases prove that financial backing does not guarantee success – and in some cases, a high budget may even fuel greater scrutiny.

Final Thought

In an era where audience sentiment is more influential than ever, studios must adapt to a world where the public – not just corporate executives – determines what succeeds or fails. The balance of power has shifted, and navigating this new reality requires engaging authentically with audiences rather than merely dictating content from the top down.

Rise of the Citizen Critic

At the heart of the new media rebellion are citizen critics and influencers who have risen through platforms like YouTube to wield substantial influence over public perception. Unlike traditional reviewers affiliated with major media outlets, these voices often come directly from the audience itself – authentic, outspoken, and fiercely independent.

Influencers like Doomcock (Overlord DVD), Critical Drinker, Nerdrotic, Geeks + Gamers, and others have become powerful opinion-makers with large and dedicated fanbases. Their critical commentary frequently challenges studio narratives, spotlighting issues with storytelling, casting decisions, and perceived political messaging in entertainment. Their blunt, often humorous, and highly relatable style resonates strongly with fans who feel increasingly disconnected from mainstream entertainment journalism.

The impact of these influencers is profound. Their ability to rapidly mobilize large online communities can significantly affect audience anticipation, viewer reception, and, ultimately, box-office performance. Examples include the amplified backlash against films like Disney's *Snow White* remake and *Indiana Jones and the Dial of Destiny*, where sustained criticism from citizen critics contributed notably to their commercial struggles.

Studios and streaming platforms now find themselves navigating a new reality: traditional marketing alone is insufficient. Engaging directly with or responding to the criticism of citizen influencers has become essential to managing public sentiment. These influencers have reshaped the power dynamics, turning everyday fans into influential critics capable of dictating media success or failure. Their rise underscores an era where community engagement, authenticity, and responsiveness are indispensable in the entertainment galaxy.

The Power of Influencers & Podcasters

The rise of independent voices in media, particularly through long-form podcasts, has given audiences new sources of information and critique. Figures like Joe Rogan, Lex Fridman, and MrBeast wield immense influence, often shaping public perception far more effectively than traditional media outlets.

The ability of these voices to drive discussion and sometimes derail projects has put studios, streaming platforms, and broadcasters on the defensive.

The Shift from 'Viral Videos' to 'Viral Judgment'

We've moved beyond the era of simple viral clips. Now, the real power lies in viral discourse, where audience sentiment itself becomes the story. This shift marks a fundamental change in how content is received:

- **Before**: A single viral video could boost a creator or project into the mainstream.
- **Now**, A wave of viral reactions (positive or negative) can determine whether a project even gets off the ground.

Navigating the New Audience-Driven Landscape

These cases highlight a profound shift in how media is received, evaluated, and ultimately judged. The emergence of ratio-ing, social media-driven backlash, and influential citizen critics underscores the growing power audiences hold in shaping media outcomes. No longer passive recipients of content, today's viewers actively influence the fate of entertainment projects, often well before their official release.

Studios and streaming platforms now face an environment where digital sentiment and community engagement directly affect commercial success. Navigating this new landscape requires transparency, authenticity, and active dialogue with audiences, acknowledging that viewer reactions, amplified through influencers, podcasts, and viral discourse, can significantly impact a project's viability.

Recognizing and adapting to this audience-driven era will be essential, not just in entertainment but across all areas of media, including live sports broadcasting, where fan engagement similarly shapes coverage, consumption, and success.

Sports Broadcasting: When Fans Hijack the Narrative

How Sports Fans Manipulate Online Discourse

Understatement of the Millenium: "*Sports fans have always been vocal.*"

Empowered by social media, fans now wield unprecedented influence in shaping narratives around games, players, teams, and broadcasters. What was once confined to local newspaper columns, radio call-in shows, or *impassioned* conversations at sports bars has evolved into a real-time global dialogue spanning platforms like Twitter, YouTube, Reddit, Instagram, Discord, and TikTok.

Fans no longer passively react after the final whistle; instead, they actively drive storylines, craft perceptions, and influence public opinion, sometimes within mere seconds of a key play or pivotal moment.

Whether through viral memes, targeted hashtags, detailed analysis videos, or even coordinated campaigns, fan voices now dictate not only how games are remembered but also how athletes and commentators are perceived, celebrated, or criticized. This new dynamic forces teams, leagues, and broadcasters to engage proactively with fan communities, recognizing that control of the narrative has shifted firmly into the hands of the audience.

Case Study: The NFL Draft and Fan-Driven Hype Cycles

The NFL Draft is a prime example of how fans hijack the narrative. Players can rise or fall in the draft order based on social media sentiment alone. Viral clips of a quarterback's perfect pass or a lineman's impressive workout numbers can generate massive fan hype, sometimes outweighing actual scouting reports.

Compound this with fantasy sports, and one bad highlight, an overthrown pass, or a fumbled snap can be clipped, memed, and shared to the point that it defines a player's reputation before they even enter the stadium.

Teams have started hiring social media analysts to monitor fan sentiment and counteract viral narratives. If a player is trending negatively, teams will flood social media with counter-narratives, emphasizing positive stats or leadership qualities. This new reality means that sports franchises aren't just managing their teams on the field. They're managing their brand in the digital arena.

Ratio-ing & Comment Wars in Sports Media

Just as films and TV shows get ratioed, sports networks and broadcasters are also subject to mass audience reactions. If a broadcaster is perceived as biased or incompetent, fans will launch online campaigns to demand their removal.

Example: ESPN and NBA Finals Coverage

During the NBA Finals, ESPN's commentary teams often face intense online scrutiny. If fans dislike a particular announcer, they take to Twitter (X) and Reddit, generating enough negative conversation that networks are forced to respond. In some cases, this has led to mid-season commentary changes or the sidelining of polarizing figures.

YouTube sports analysts and independent commentators have capitalized on this, offering alternative coverage that is often more popular than network broadcasts. Some fans now prefer watching a game while listening to their favorite YouTuber or podcaster break it down in real time, bypassing traditional broadcasters altogether.

The Power of Fan-Led Podcasts and Independent Analysis

The rise of independent sports commentators has further disrupted the industry. Once upon a time, the only voices that mattered were those of major network analysts. Now, fan-led podcasts, YouTube breakdowns, and Twitter (X) spaces can generate more engagement than traditional media outlets.

- The Pat McAfee Show, with nearly 2.8M followers at the time of this writing, for example, has become one of the most influential sports programs despite not being tied to a major network.
- Analysts on YouTube, such as Brett Kollmann and Jomboy Media, offer breakdowns that rival (or surpass) network coverage.
- Platforms like Discord and Reddit allow fans to create their own game analysis hubs, sometimes more in-depth than ESPN's coverage.

The future of sports broadcasting will not be controlled solely by networks. It will be a blend of mainstream coverage, independent voices, and real-time audience engagement.

The Impact of AI & Algorithmic Sports Coverage

The Rise of AI-Generated Highlights

AI is changing how sports are consumed. In the past, fans had to wait for networks to produce highlight reels. Now, AI can generate personalized highlight packages in real time, cutting together key plays based on a user's preferences.

Example: The NBA's Automated Highlight System

The NBA has partnered with AI companies to automatically generate highlights for fans. Instead of watching a full game, a user can receive an AI-curated package featuring only three-pointers, dunks, or defensive plays tailored to their interests.

NBA's AI-Driven Personalized Highlights

Cleveland Cavaliers' Use of AI: Since 2015, the Cleveland Cavaliers have partnered with WSC Sports' AI platform to automatically create and distribute real-time game highlights. This system enables the generation of personalized highlight reels, allowing fans to receive content that aligns with their specific interests. https://www.youtube.com/watch?v=3NKjoKvFwAQ

NBA's Automated Video Clipping Solution: The NBA has implemented an automated video clipping solution that generates real-time, customizable content. This system analyzes various data points, including on-screen graphics and audio cues, to determine optimal moments for highlight creation. Such technology allows fans to access tailored highlight packages featuring specific types of plays or favorite players. https://www.sportsvideo.org/2015/12/07/nba-digitals-new-automated-video-clipping-solution-generates-real-time-customizable-content/

Impact on Fan Engagement

By leveraging AI to deliver personalized highlight packages, the NBA enhances the fan experience, offering content that caters to individual preferences. This approach not only increases engagement but also allows fans to interact with the sport on their terms, selecting the content that resonates most with them.

The integration of AI in generating personalized sports highlights exemplifies the evolving landscape of sports consumption, where technology plays a pivotal role in delivering customized and engaging content to fans.

For a visual demonstration of how AI is utilized in creating personalized video content, you can watch the following video: https://youtu.be/3NKjoKvFwAQ

This raises a major question:

If fans can consume sports through AI-curated highlights, do they need traditional broadcasts at all?

Will AI Replace Sports Broadcasters?

AI-powered commentary is also emerging. Some platforms are experimenting with machine-generated play-by-play analysis, using deep learning models trained on thousands of hours of sports coverage.

Artificial Intelligence (AI) is making significant inroads into sports broadcasting, particularly in the realm of live commentary. By leveraging deep learning models trained on extensive datasets, AI systems are now capable of generating real-time play-by-play analyses. Notable examples include:

Bundesliga's AI-Powered Live Commentary

The German Bundesliga, in collaboration with Amazon Web Services (AWS), has implemented a generative AI solution that produces real-time, automated commentaries during matches. This system analyzes over 1,600 events per game – such as goals, passes, and fouls – to generate commentary in multiple languages and styles within seconds, enhancing accessibility and personalization for fans worldwide. https://aws.amazon.com/blogs/media/revolutionizing-fan-engagementcer-bundesliga-generative-ai-powered-live-commentary/,https://www.rockingrobots.com/ai-provides-live-sports-commentary/

IBM's AI Commentary at Major Tennis Tournaments

IBM has introduced an AI commentary system that is deployed at events like Wimbledon and the U.S. Open. This technology utilizes generative AI to provide automated narrations for highlight packages, offering fans insights and summaries without human intervention. https://arxiv.org/abs/2402.15514

Emerging AI Commentary Systems

Innovators are developing AI-driven commentary systems capable of real-time analysis and multilingual support. For instance, a developer showcased an AI sports commentator that detects key moments in videos and provides commentary without manual prompts, utilizing tools like Whisper for speech-to-text, Gemini Flash for vision processing, and ElevenLabs for voice synthesis. https://www.reddit.com/r/MachineLearning/comments/1fnry1x/p_i_built_a_live_ai_sports_commentator_that_can/

These advancements indicate a trend towards integrating AI in sports broadcasting, offering personalized and immediate commentary experiences. While AI-generated commentary enhances accessibility and provides novel viewing options, it also raises questions about the future role of human broadcasters in delivering nuanced and emotionally resonant analyses.

For a visual demonstration of AI in sports commentary, you might find the following video insightful:… https://youtu.be/u56K4dL20gA

Pros and Cons:

- Pros: AI can offer unbiased, data-driven insights in real-time, adapting to viewer preferences.

- Cons: Lacks the emotional connection and storytelling that human broadcasters provide.

While AI may never fully replace live commentators, it is likely to reshape how sports are presented. In the near future, fans might have the option to toggle between human and AI commentators or even generate custom, hyper-personalized broadcasts.

The Future of Live Sports Streaming: Personalized Feeds and Betting Integration

As AI and engagement algorithms evolve, we are approaching a future where sports broadcasts are completely customizable.

- Fans could choose different camera angles, commentary teams, or statistical overlays.
- Real-time sports betting integration will allow viewers to place bets within the stream itself.
- Augmented reality (AR) could bring real-time stats, player tracking, and interactive experiences to live games.

The way we consume sports is evolving at an unprecedented pace. The traditional, one-size-fits-all broadcast is dying, replaced by a fragmented, user-driven experience. Sports leagues and networks must adapt or risk losing control of their content to independent creators, AI-generated coverage, and fan-led discourse.

The next section will explore whether platforms can regain control or if the audience has permanently taken the reins of media consumption.

The Future: Can Platforms Regain Control?

Will Streaming Giants Start Fighting Back?

The rise of user-driven influence through ratio-ing, influencer commentary, and algorithmic hijacking has put platforms and media corporations in an unfamiliar position: reacting instead of leading. The question now is whether they will find ways to regain control or if the power shift is permanent.

One of the first responses from major platforms has been to reduce the visibility of audience feedback. YouTube has already removed public dislike counts on videos to prevent mass negative engagement campaigns. Rotten Tomatoes has implemented stricter review controls to combat early review bombing of films and TV shows. Amazon took a similar approach when it restricted user reviews on The Rings of Power after backlash. These moves suggest that platforms recognize the influence of mass audience reactions and are actively working to limit their impact.

However, these strategies come with risks. Suppressing audience input can backfire, as users see it as censorship or manipulation, further fueling distrust in platforms. The removal of public dislike counts on YouTube, for example, led many users to install third-party browser extensions to restore them, demonstrating that audiences will find ways to take back control.

The Shift Toward Walled Gardens and Subscription Models

Another likely response from platforms is to revisit the strategy of walled gardens and paywalls – a tactic audiences have seen repeatedly rise, fall, and resurface. This isn't new territory; cable TV packages once dominated through bundling, and early internet giants like AOL thrived by corralling users inside controlled ecosystems. Now, platforms like Disney+ merging Hulu and ESPN+, or Netflix venturing into gaming, are attempting similar moves, betting on audiences preferring curated environments rather than the chaotic freedom of social media.

Yet history shows audiences inevitably resist these cycles. Each attempt at tighter control eventually gives way to new disruptions, alternative channels, and independent voices. The question isn't just whether audiences will accept these renewed paywalls but how long they'll tolerate them before seeking or creating new avenues of engagement.

What Happens Next?

The future of content distribution remains more uncertain than ever. On the one hand, streaming platforms and media corporations will continue refining their algorithms, restricting audience manipulation, and attempting to maintain control. On the other, audiences have never been more empowered to dictate which content rises and which is buried.

Emerging technologies like decentralized media platforms, blockchain-based content verification, and AI-driven personalized feeds may further disrupt traditional broadcasting and streaming models. We could see new platforms where audience voting directly influences rankings rather than hidden recommendation algorithms deciding for them.

At the same time, the influence of independent creators, podcasters, and fan-driven analysis is only growing. Traditional media outlets may no longer be the primary source of authority on entertainment, sports, and culture. Instead, we could see an era where content is curated more by communities than by corporations. At least, that is, until the cycle repeats, and the next-gen of content control tools supersedes those communities again.

Conclusion: The Battle Between Algorithms & Audience Influence

The balance of power is shifting once more in the entertainment galaxy. What began as a rebellion against the empire of TV executives and studio gatekeepers evolved into a new battle against algorithmic control wielded by the powerful streaming platforms. Now, audiences have discovered their own Force, pushing back through ratio-ing, podcasters, citizen critics, and influencer-driven campaigns, reshaping what succeeds and what fails.

Yet, as always, the Empire seeks ways to regain dominance – restricting audience feedback, tightening control within walled gardens, and harnessing AI to customize and constrain viewer experiences. But audiences are relentless, continually finding new ways to disrupt, criticize, and influence.

Who ultimately controls our screens? Is it the mighty platforms with their data-driven, algorithmic Death Stars, or is it the passionate, vocal Rebel audience fighting for the content they truly value? The future remains uncertain, but one truth emerges clearly: The struggle between the algorithmic Empire and audience rebellion will define the next chapter of digital entertainment.

In this ongoing Galactic Civil War of media, neither AI nor executives hold absolute power. The audience – armed with engagement tactics and independent voices – has become a critical part of the equation. Whether platforms embrace this change or fight to suppress it will determine the fate of entertainment itself. As we look ahead, the Force – viewer choice and engagement – may finally be poised to truly awaken.

The fight isn't over. Stay informed, stay engaged, and make your voice heard. Share your thoughts, challenge the algorithm, and be part of the conversation. The future of entertainment belongs to those who shape it.

14: Beyond Broadcast.com

Tackling Stadium Connectivity Challenges Today

B ack in the early days of AudioNet.com – later Broadcast.com – streaming a live game typically meant one of two things. We either pulled the audio feed from a radio station we already carried or we relied on a satellite or point-to-point feed we'd set up for that specific broadcast. Once we had that raw signal, we encoded the audio and sent it across our servers as packets to fans listening through dial-up connections and, later, DSL. In truth, that fundamental process hasn't changed as much as you might think. Today, whether you're streaming a game from a smartphone, a smart TV, or a laptop, it's still a matter of pulling down a feed, encoding it, and sending out those packets to end users all over the world.

Of course, what has changed is how we consume those streams and our expectations for quality. In the '90s, simply hearing the play-by-play on your computer felt magical – even if it was a bit choppy or delayed. Now, we expect high-fidelity, near-instant audio and video on devices that fit in our pockets. Yet the same basic mechanics remain in play: your stream depends on packets arriving, being reassembled, and then playing back in proper sequence. That's how services maintain quality. When a packet is late, the stream often pauses or buffers so it can "catch up," creating delays that can run anywhere from half a minute to several minutes behind real-time action.

This built-in latency was acceptable (or at least tolerable) when we were tethered to desktop computers or waiting for a still image to load. But in a stadium where everything is happening right in front of you – and you might want to rewind or watch replays on your phone – the lag becomes glaringly obvious. Stadium connectivity only compounds the challenge, as thousands of fans all try to stream and share simultaneously.

In this chapter, we'll explore why, decades after the early days of AudioNet.com / Broadcast.com, stadiums still struggle to keep fans connected, the technology behind these bottlenecks, and how sports venues might finally solve the connectivity puzzle once and for all.

The Growing Role of Phones in Stadiums

In today's sports arenas, smartphones have become integral to the fan experience, enabling real-time engagement through activities such as checking live statistics, watching replays, participating in live betting, and sharing moments on social media. Yet, latency becomes a significant issue not only for fans attending events but also for those streaming remotely while interacting with friends via text, FaceTime, or social media. As one fan pointed out, even slight delays in the stream can introduce a frustrating "spoiler effect" among groups watching virtually, as individuals receive feeds at slightly different times. A friend texting a reaction to a critical play seconds before others see it themselves can ruin the immediacy and excitement that makes the live sports experience uniquely compelling.

By the Numbers:

Prevalence of Smartphone Use:

A study by the University of Florida found that the convenience of smartphone sports consumption has helped the platform compete for fan attention with other media outlets such as television. https://www.jou.ufl.edu/insights/smart-phones-have-significant-impact-on-sports-fans-behavior/

Changing Fan Expectations:

Modern fans anticipate seamless connectivity within stadiums, expecting services like mobile food ordering and instant access to highlights. However, many venues struggle to meet these expectations due to connectivity challenges. https://edtechmagazine.com/higher/article/2022/08/how-mobile-technology-impacts-college-sports-fan-experience

Connectivity Challenges in Stadiums:

Despite the critical role of smartphones in enhancing the live sports experience, stadiums often face difficulties in providing reliable mobile connectivity. The high concentration of users can overwhelm networks, leading to slow speeds and inconsistent service. Factors such as the large number of simultaneous connections and the physical structure of stadiums contribute to these issues. https://engagemintpartners.com/chatting-sports-tech-edition-13-in-stadium-connectivity/

To address these challenges, some venues are investing in advanced Wi-Fi infrastructure and exploring technologies like 5G to improve connectivity and meet fan expectations. https://www.verizon.com/business/resources/articles/s/the-smart-stadium-of-the-future/

As smartphones continue to play a pivotal role in how fans interact with live events, enhancing in-stadium connectivity remains essential for delivering a satisfying and immersive fan experience.

Why AM/FM, WiFi, and Traditional Streams Struggle Inside Stadiums

In the context of modern stadiums, traditional methods of accessing game commentary and content, such as AM/

FM radio and standard Wi-Fi or streaming services, often fall short of meeting contemporary fan expectations.

AM/FM Radio: A Relic of the Past

- **Continued Usage:** Some fans still bring portable radios to games to access real-time commentary, as this method is unaffected by potential Wi-Fi or cellular connectivity issues within crowded stadiums.
- **Limitations:** AM/FM radio lacks interactivity and flexibility, offering no personalized or on-demand content. In an era where fans desire instant replays, customizable updates, and interactive features, traditional radio fails to meet these evolving expectations. Additionally, the static nature of AM/FM broadcasts does not align with the dynamic, immersive experiences that modern fans seek during live events. Moreover, in today's multi-tiered and complex stadium architectures, AM/FM signals are often blocked and suffer from dead reception zones.
- As fan engagement continues to evolve, there is a growing demand for more interactive and personalized content delivery methods that traditional AM/FM radio cannot provide.

Public WiFi in Stadiums

Public Wi-Fi networks in stadiums are implemented to enhance the fan experience by providing internet connectivity during events. However, these systems often encounter several challenges:

The Reality:

- **Congestion:** With thousands of fans attempting to connect simultaneously, stadium Wi-Fi networks can become overwhelmed, leading to slow speeds and dropped connections. For instance, during a game at the University of Florida's Ben Hill Griffin Stadium, over 56,000 unique devices connected to the Wi-Fi network, transferring 13.6 terabytes of data, illustrating the immense demand placed on such systems. https://edtechmagazine.com/higher/article/2023/05/upgrading-stadium-wi-fi-connects-fans-their-universities

- **Coverage Gaps:** The vast size and complex architecture of stadiums, often constructed with extensive concrete and metal, can impede wireless signals, resulting in uneven coverage and areas without access. Older stadiums, in particular, may require significant modifications to accommodate comprehensive Wi-Fi networks, including drilling through thick concrete to install access points. https://www.sportsbusinessjournal.com/Journal/Issues/2017/03/20/In-Depth/College-wifi.aspx

- **Security Concerns:** Public Wi-Fi networks are often unsecured, exposing users to potential data breaches and cyber threats and venues to potential legal liabilities. The high concentration of users in stadiums can make these networks attractive targets for malicious activities, necessitating robust security measures to protect fan data. https://msb-engineering.com/we-connect/why-stadiums-and-arenas-need-wifi-6/

Addressing these challenges requires substantial investment in advanced networking technologies, such as Wi-Fi 6, which offers improved capacity and efficiency for high-density environments like stadiums. Additionally, implementing comprehensive security protocols is essential to safeguard user data and maintain trust. https://www.wi-fi.org/file/wi-fi-6-in-stadiums

Traditional Streaming Services

Traditional streaming services, such as ESPN+ and YouTube TV, often encounter significant challenges when utilized within stadium environments:

Latency Issues

These platforms typically experience delays ranging from 45 to 60 seconds behind real-time action. This latency can disrupt the in-person experience for fans attempting to view replays or access alternate camera angles during live events. https://www.mediakind. com/blog/2022-the-year-live-sports-streaming-kicks-latency-issues-into-touch/

Bandwidth Constraints

In crowded stadiums, the simultaneous use of cellular networks by thousands of fans can lead to network congestion. This high demand often results in buffering, lag, or complete service interruptions, as the available bandwidth becomes insufficient to support extensive streaming activities. https://thelastmile.gotennapro.com/ can-stadium-security-teams-sidestep-congested-networks/

These challenges highlight the limitations of traditional streaming services in delivering a seamless and timely viewing experience within densely populated venues.

The Challenges of Stadium Connectivity

Providing seamless connectivity in stadiums presents significant engineering and logistical challenges due to several factors:

High User Density

Bandwidth Demand: A sold-out stadium can host tens of thousands of attendees, each utilizing multiple devices to access the internet simultaneously. Traditional cellular towers and Wi-Fi networks are often not designed to accommodate such concentrated demand, leading to network congestion and degraded performance.https://www.fs.com/blog/things-you-should-know-about-wifi-for-large-stadiums-and-venues-2807.html

Latency

Real-Time Requirements: Fans increasingly engage in activities that require real-time connectivity, such as live betting, streaming replays, and accessing interactive content. High traffic volumes and network bottlenecks can introduce delays, frustrating users who expect instantaneous responses, especially during fast-paced sports events. https://networkbuilders.intel.com/docs/networkbuilders/enhancing-the-fan-experience-optimizing-operations-intel-1718264110.pdf

Physical Barriers

Architectural Obstacles: Stadium structures, composed of materials like concrete and steel, can obstruct wireless signals, creating coverage gaps and inconsistent connectivity. The complex layouts and large areas involved make it challenging to ensure uniform signal distribution throughout the venue. https://www.networkcomputing.com/network-management/designing-connected-stadiums-for-better-fan-experiences-and-streamlined-operations

Evolving Expectations

Demand for Advanced Connectivity: Modern fans expect more than basic internet access; they desire high-speed, low-latency networks that support immersive experiences, such as augmented reality (AR) features, live polls, and interactive applications. Meeting these evolving expectations requires continuous investment in advanced networking technologies and infrastructure. https://www.wi-fi.org/beacon/dennis-huang/wi-fi-6-bolsters-the-stadium-fan-experience

Addressing these challenges is crucial for enhancing the fan experience and ensuring that stadiums can meet the connectivity demands of contemporary audiences.

What Needs to Change: Solving the Connectivity Problem

Enhancing connectivity in stadiums is essential to meet the growing demands of fans for seamless, high-speed, and interactive experiences. Implementing advanced technologies can address current challenges and significantly improve network performance.

Distributed Antenna Systems (DAS)

How It Works: DAS involves deploying a network of spatially separated antennas throughout a stadium to enhance cellular coverage and capacity. These antennas distribute the cellular signal, reducing the load on any single point and mitigating congestion issues.

The Challenge: Implementing DAS is complex and requires substantial investment. It necessitates collaboration among network providers, venue operators, and equipment manufacturers to ensure seamless integration and operation. Additionally, designing a DAS that accommodates a venue's maximum capacity, even though such capacity is only reached during a few events annually, poses a financial challenge for venue owners. https://piersonwireless.com/distributed-antenna-systems/stadium-das-challenge/

Private 5G Networks

Why It's Promising: Private 5G networks offer ultra-low latency, high-speed data transfer, and the capacity to support thousands of simultaneous connections. They provide dedicated bandwidth, ensuring reliable connectivity even during peak usage.

Examples: Venues like the Stade Vélodrome in Marseille have integrated private 5G networks to enhance the fan experience. These networks support applications such as real-time video streaming, biometric access control, and efficient operations management. https://

www.datacenterdynamics.com/en/analysis/private-5g-network-orange-stade-velodrome/

Edge Computing

The Solution: Edge computing processes data closer to the user by utilizing localized servers, thereby reducing latency and improving responsiveness for live applications like streaming and in-stadium augmented reality (AR).

Applications: By processing data at the network's edge, stadiums can offer features like instant replays, live betting updates, and seamless mobile ticketing without overwhelming central servers. For instance, during live events, edge computing can facilitate real-time analytics and personalized content delivery to fans' devices. https://itcblogs.currentanalysis.com/2022/08/22/edge-computing-is-enabling-new-applications-in-stadiums-and-other-large-venues/

Hybrid Connectivity Solutions

Combining Technologies: Integrating public Wi-Fi, private 5G, and edge computing can create robust networks that adapt to varying demands during events. This hybrid approach ensures that different applications receive the appropriate level of connectivity and bandwidth.

Implementation: For example, a stadium might use private 5G networks to support mission-critical operations and high-bandwidth applications, while public Wi-Fi can handle general internet access for fans. Edge computing can be employed to manage data processing for time-sensitive applications, ensuring a seamless and responsive experience. https://stlpartners.com/articles/private-cellular/private-networks-in-hospitality-and-venues/

By adopting these advanced connectivity solutions, stadiums can overcome existing challenges and provide fans with a more immersive and reliable experience during events.

The Stakes for Stadiums and Sports Organizations

Addressing connectivity issues in stadiums is not merely a technical concern but a critical business imperative for sports organizations. Inadequate connectivity can lead to significant negative outcomes, while effective solutions can enhance fan engagement and unlock new revenue streams.

Consequences of Poor Connectivity

- **Declining Attendance:** Modern fans expect seamless digital experiences during live events. A lack of reliable connectivity can deter attendance, as fans may prefer the comfort and technological advantages of home viewing. Studies have shown that despite increasing revenues, stadium attendance has stagnated or decreased slightly across many leagues, indicating that fans may opt to stay home if their in-venue expectations are not met. https://www2.deloitte.com/content/dam/Deloitte/us/Documents/technology-media-telecommunications/us-tmt-stadium-experience-keeping-sports-fans-engaged-loyal.pdf

- **Lost Revenue Opportunities:** Reliable connectivity is essential for in-stadium services such as mobile concessions, merchandise sales, and live betting. A survey revealed that 81% of event-goers would spend more if there were more convenient food, beverage, or merchandise ordering options, and 57% would spend more if they could order food and beverages from their seats. Without robust connectivity, these potential revenue streams remain underutilized. https://www.boldyn.com/us/news/the-future-of-fan-experience-belongs-to-truly-connected-venues-study-reveals

Benefits of Enhanced Connectivity

- **Enhanced Engagement:** Providing personalized and connected experiences fosters fan loyalty and immersion. Features such as interactive displays, real-time statistics, and dynamic content can significantly improve the fan experience, making it more enjoyable and immersive. https://www.sportsvenue-technology.com/ articles/smart-stadiums-connectivity-and-iot-solutions-for-fan-engagement

- **New Revenue Streams:** Improved connectivity enables innovative features like augmented reality (AR) overlays, interactive statistics, and targeted advertising. For instance, 72% of event-goers expressed interest in accessing real-time player statistics through AR overlays, presenting opportunities for monetization through sponsorships and premium services. https://www.boldyn.com/us/news/the-future-of-fan-experience-belongs-to-truly-connected-venues-study-reveals

Investing in robust connectivity infrastructure is essential for stadiums and sports organizations aiming to meet modern fan expectations and capitalize on emerging revenue opportunities.

Conclusion: A Connected Future for Stadiums

As we've seen throughout this series – from the early days of AudioNet/Broadcast.com through the on-demand revolution – innovations in connectivity and content delivery have always been about bridging distances. The same impulse that once drove us to climb rooftops and hook up homemade encoders has now transformed into the global race for stadium-wide connectivity solutions. Fans may no longer be at home on dial-up modems waiting for a live feed to buffer, but they are in the stands expecting instant replays, real-time stats, augmented reality overlays, and interactive social experiences – all of which demand robust, low-latency networks.

This evolution highlights a recurring lesson from previous chapters: it's never just about the technology. Content is still king, and "access" has replaced "ownership" across nearly every part of the media ecosystem. The challenge for stadiums is to provide an immersive experience that's better than, or at least complementary to, what fans can get in their living rooms. That means embracing private 5G networks, distributed antenna systems (DAS), edge computing, and other advanced solutions that ensure streams aren't just available – they're seamless and rewarding.

More than ever, this is a business imperative. Fans expect frictionless connectivity whether they're checking fantasy stats, posting on social media, or watching replays of the play they just witnessed on the field. Neglecting these demands can drive people back to the comfort of their couches, where "buffering" has been replaced with "bingeing." Stadiums that fail to evolve risk becoming relics in an era defined by on-demand media and interactive digital experiences.

Yet, as the prior chapters remind us, adopting new technology also raises questions of control, curation, and who ultimately shapes the fan experience. In the same way that streaming redefined how we consume music and video at home, modern connectivity is redefining what it means to attend a live event. We stand at the intersection of legacy sports traditions and next-generation digital engagement, and the choices stadiums make will set the tone for how future generations experience "live" sports and entertainment.

The question is no longer whether stadiums can deliver connectivity – they must. What remains to be seen is how quickly (and how wisely) they can adapt to provide the high-speed, interactive experiences fans now consider essential. As the series continues, we'll look at how these connectivity struggles echo the broader, ongoing shifts in our media landscape – from physical to digital, from owning to streaming, and from being passive spectators to active participants. After all, the next big leap in fan engagement might just be waiting for the stadium that gets it right first.

What innovations do you think will transform the in-stadium experience?

15: Broadcast.com's Legacy

and the Last Frontier of Live Sports Streaming

1. Introduction: Why This Still Isn't Solved

Live sports remain one of the last bastions of true communal entertainment. Stadiums roar with energy, broadcasters relay the action to millions, and fans worldwide stay glued to their screens. But even in this connected age, one problem lingers: delivering real-time, high-quality color commentary to fans both in the stadium and around the globe.

When we built Broadcast.com in the '90s, delivering a live game over the internet was a marvel. To accomplish this, we had to do one of three things. We either pulled the feed from a station we already carried, relied on a satellite or point-to-point feed we'd set up for that specific broadcast, or received the feed via a direct telco codec connection. Once we had that raw signal, we encoded the audio and sent it across our servers as packets to fans listening through dial-up connections and, later, DSL.

Of course, I doubt anyone today is standing in the corner of a dusty repurposed bachelor pad, next to a humming 19" rack, listening to a dozen feeds at once, attuned to spotting the slightest dropout or crackle. Back then, that was the job, using your ears as the quality control system. Today, we've got software that does much of that for us, with real-time diagnostics and even algorithms trained to detect issues in video and audio streams before a human would notice.

Today, streaming has become seamless and beautiful, until you try to use it in a stadium. Despite enormous progress in infrastructure and tools, fans in the stands are often left in a kind of digital silence. They see the play unfold live, but if they try to listen to their favorite color commentator at the same time, it will lag thirty seconds, or even up to a few minutes delayed, and by then you may have forgotten what they're even talking about.

We've come a long way in terms of tools and automation, but in other ways, the game still feels familiar. just with fewer ethernet cables dangling from the ceiling like so many jungle vines or cheap cafeteria tables turned into improvised server racks bowed in the middle from the weight, and way more KPIs and dashboards.

In truth, that fundamental process hasn't changed as much as you might think. Today, whether you're streaming a game from a smartphone, a smart TV, or a laptop, it's still a matter of pulling down a feed, encoding it, and sending out those packets to end users all over the world.

Of course, what has changed is how we consume those streams and our expectations for quality. In the '90s, simply hearing the play-by-play on your computer felt magical -- even if it was a bit choppy or delayed. Now, we expect high-fidelity, near-instant audio and video on devices that fit in our pockets. Yet the same basic mechanics remain in play: your stream depends on packets arriving, being reassembled, and then playing back in proper sequence. That's how services maintain quality. When a packet is late, the stream often pauses or buffers so it can "catch up," creating delays that can run anywhere from half a minute to several minutes behind real-time action.

This built-in latency was acceptable (or at least tolerable) when we were tethered to desktop computers or waiting for a still image to load. But in a stadium where everything is happening right in front of you

-- and you might want to rewind or watch replays on your phone -- the lag becomes glaringly obvious. Stadium connectivity only compounds the challenge, as tens of thousands of fans all try to text, talk, surf social media, consume streams and share with their friends simultaneously.

In-stadium streaming isn't just about convenience; it's about enriching the entire experience. Imagine being at a Cowboys game and hearing Tony Romo's analysis in real-time as the play unfolds before your eyes. Or picture yourself at a Premier League match, choosing commentary in your native language while the crowd around you erupts. These aren't just nice-to-have features anymore. They're the missing pieces in a puzzle we've been trying to solve since the early days of Broadcast.com.

The stakes couldn't be higher. As stadium attendance faces competition from increasingly sophisticated home viewing setups, venues need every advantage to get fans through the turnstiles. Teams, leagues and sponsors are looking for new revenue streams beyond traditional broadcast rights. And fans. who are paying more than ever for tickets. are demanding experiences that justify the premium pricing.

This chapter isn't just a look at the technology of low-latency audio. It's an examination of why, even after three decades of innovation, we still haven't cracked one of the most basic and powerful use cases in sports: giving fans at the game the same commentary and context that viewers at home take for granted.

And while companies are getting close, the problem isn't just one of bandwidth or signal. It's about architecture, user experience, business model, and venue politics. That's why this isn't solved. Yet.

In this chapter, we'll explore what's changed since the early days of Broadcast.com, what hasn't, and why real-time audio might be the missing piece that redefines the stadium experience for fans, teams, and broadcasters alike.

2. The Fan Experience Gap

Go to any game these days, and you'll see the same thing: fans juggling between watching the action on the field and checking their phones for replays, stats, betting updates, fantasy scores, commentary, (or Twitter/X or TikTok or Facebook, or….). They're trying to stitch together their own second-screen experience in real time, but the tools to do that simply aren't there. Not yet.

In-stadium silence is a very real thing. Sure, you've got the roar of the crowd, a halftime show, a jumbotron blast of hype, and maybe a PA announcer (if you can hear it over the crowd), but the insightful commentary that enriches broadcasts is missing for the in-person fan. Home viewers get expert analysis, instant replays, and play-by-play breakdowns. In the stands, you might not even hear what the penalty was. You're close to the action, but detached from its meaning.

Fans try to bring earbuds and tune in to the radio feed, but that often ends in frustration. The AM/FM signal can't penetrate modern stadium construction. The stream is delayed by 30 to 60 seconds or more. So instead of enriching the moment, it shatters it. You hear the excitement over a play that already happened. Meanwhile, the next one is already unfolding right in front of you.

International fans face another layer of friction. A Japanese NBA fan at a Lakers game, a Brazilian soccer enthusiast at a Champions League match, they're in the stadium, living the dream. But they can't follow the English-language commentary or cultural shorthand. There's no alternative audio, no real-time translation, no effort made to bring them into the moment. The result is that they often experience the event as pure spectacle, not story.

And then there are fantasy players, who now number over 60 million in the U.S. alone, and sports bettors, two segments leagues are heavily invested in courting. These fans are deeply engaged. They need information now, not after the fact. Yet their in-venue experience is often worse than what they get at home.

The irony is stark. We've spent the last two decades making the at-home experience better, ultra-HD streams, multiple camera angles, live stats, interactive overlays, and the in-stadium experience, technologically, hasn't kept pace. Sure, there are better seats and better food. But the data, the insight, the context that fans have grown to expect simply isn't there.

And that's the gap. It's not just about signal strength or latency. It's about emotional presence. It's about bringing fans closer to the story, in the moment, without delay, and without exclusion, not to mention justifying the cost and effort of attending games in person.

Solving that problem is the next evolution of the stadium experience. And it starts with audio.

The stakes go beyond bandwidth and protocols; they touch on the very future of live sports attendance in a digital world.

---

3. The Business Case for Real-Time Audio$

Let's be clear, real-time audio in stadiums isn't just a cool idea for fans. It's a business lever hiding in plain sight.

When done right, real-time audio isn't just about commentary. It's about engagement. It's about data. It's about money. And it just might be the most cost-effective way to bridge the growing gap between physical attendance and the digital expectations of modern fans.

Start with fan satisfaction. Deployments of real-time audio have shown a measurable lift in Net Promoter Scores, with fans who use in-venue commentary reporting higher satisfaction and a greater likelihood to return.

One of the strongest drivers behind real-time audio deployment is the opportunity to create deeper fan engagement. Modern audiences, particularly Gen Z and Millennials, expect continuous context and commentary, even during live play. A 2023 Sports Innovation Lab study found that 68% of these fans check their phones for updates at least 10 times per game. Real-time audio satisfies this craving for information while keeping eyes on the field.

In-stadium deployments consistently show:

- +12–18 Net Promoter Score gains among real-time audio users

- +22 minutes longer average stay in the stadium
- +34% higher return rate for attendees who used audio services
- +41% increase in social sharing from users during games

By restoring access to commentary, whether traditional, personality-driven, or multilingual, real-time audio helps fans stay connected to the moment and feel more immersed in the experience.

Now layer in the monetization. Real-time audio can carry sponsor reads, branded segments, or team-centric promotions. Want to push a merch offer or upsell a premium experience? It is far more effective in someone's ears than on a banner ad lost in the noise. Teams can also offer premium audio tiers, alternate feeds, player mic access, behind-the-scenes segments, as part of VIP packages or app upgrades.

In some deployments, team-branded headphones or earpieces are offered as part of premium seating packages, fan loyalty programs, or sold as stand-alone merchandise. These physical items serve a dual purpose, as functional tools for the in-venue audio experience and as branded memorabilia. For teams and leagues, it creates a hybrid physical-digital revenue stream that deepens fan loyalty while opening new sponsor integration opportunities.

The results are measurable. In one deployment, fans who used in-venue audio were 34% more likely to return for future games. Sponsor impressions tied to audio yielded strong brand recall and higher engagement, with no additional screen time or venue footprint required.

And don't underestimate the data. Every time a fan selects a feed, toggles languages, or listens longer than average, you are capturing preferences. That behavioral data can feed CRM systems, drive smarter campaigns, and improve everything from concessions to content programming.

When a fan chooses a specific feed, home vs. away, betting-focused, Spanish-language, they're expressing preference and intent. That data, when tied to first-party identity through app engagement, becomes part of a broader CRM and fan intelligence ecosystem. It allows teams and leagues to refine marketing efforts, improve segmentation, and offer more valuable insights to sponsors and partners.

Here's another key point. Teams and leagues are desperate to get fans to use their apps. That's where the loyalty programs live, where the ticketing data is tracked, where the sponsors are waiting. But too often, those apps are just scanned once for entry, then ignored. Real-time audio gives fans a reason to open the app and keep it open. It becomes the gateway to a richer in-venue experience, not just a barcode on your phone.

Then there is the accessibility upside. In a time when stadiums are under pressure to deliver equitable experiences, real-time audio with multi-language options or descriptive services can meet both ADA and global accessibility standards. That is not just feel-good compliance. It is brand enhancement, and in some jurisdictions, a requirement.

Best of all, it doesn't require a massive infrastructure overhaul. Many platforms are BYOD, bring your own device. No headsets to rent, no proprietary hardware to manage. Just a QR code, a tap, and suddenly your fans are hearing the game in sync, in their language, in their seat.

Real-time audio solves a very real fan pain point, enhances inclusivity, drives adoption of first-party apps, and opens new revenue streams. All without asking fans to do anything they are not already doing.

Real-time audio is not just a user feature.
It's a business case.

---

4. The Technical Landscape and Approaches

When people think of streaming challenges, they often focus on bandwidth. But when it comes to real-time audio in stadiums, the challenge is less about raw speed and more about architecture, synchronization, and environment. The underlying technology stack and delivery model make all the difference.

There are a handful of primary approaches to solving the real-time audio problem, each with tradeoffs in latency, scale, complexity, and user experience. The examples provided below are not intended to be exhaustive but rather illustrative of each approach. My inclusion or exclusion of any particular company should not be construed as an endorsement or criticism. All information referenced is based on publicly available data from company websites, press releases, and industry publications as of this writing. Different solutions may have evolved since this assessment, and I encourage readers to conduct their own research on current offerings.

1. WebRTC-Based Systems

WebRTC was built for low-latency, peer-to-peer communication, and it remains the gold standard for delivering sub-second streams. Companies like Phenix and Red5 Pro use variations of this protocol to achieve latencies under 500 milliseconds at scale. But it is not magic. Delivering WebRTC to thousands of concurrent users in a stadium environment requires a distributed mesh or custom infrastructure which can get technically expensive and complex.

2. App-Based Local Wi-Fi Streaming

Mixhalo and similar players leverage the venue's own Wi-Fi to create localized delivery zones. Audio is routed over the internal network with latencies as low as 100 milliseconds. It is fast, but it depends entirely on the strength and stability of the stadium's infrastructure.

3. Browser-Based Audio Delivery

Prowire's model is lightweight. Fans access audio through a mobile browser, no app install, no login. It works over either Wi-Fi or cellular and uses smart pre-buffering to stay synchronized. But because it relies on public networks, performance may vary. Interactivity or customization can be problematic. What you gain in simplicity, you may lose in richness.

4. RF and Hybrid Systems

Some companies, like DVOX, blend old and new. They use localized RF transmitters to create micro-broadcast zones inside the venue, paired with software to manage sync and personalization. This approach sidesteps congestion entirely but requires physical installation, hardware on-site, pre-event calibration, and coordination with venue ops. It scales well once in place, but the up-front lift is higher.

5. Cloud-Driven Infrastructure

Then there are the companies sitting deeper in the stack, Dolby. io, Agora, Wowza, Harmonic. These are toolkits and platforms, not fan-facing products. They provide encoding, transcoding, and delivery services that power others. The flexibility is enormous, but it comes with the expectation that you will build and integrate your own solution.

Each of these technical choices reflects a deeper strategic one: who controls the experience, who owns the data, and how much of the value chain a team or venue wants to touch.

There is no single winner yet. But what is clear is that the future favors approaches that are:

- Low latency without massive overhead
- Easy for fans to access and understand
- Scalable without over-reliance on perfect infrastructure
- Flexibility and fast setup
- Compatible with first-party apps and existing ticketing or loyalty systems

This is not just a technical race. It is a usability contest. And in that race, whoever can deliver real-time audio that "just works," without forcing a download, a login, or a learning curve, is going to win.

At broadcast.com we treated it like a broadcast and got it done. Whoever wins will need to do the same and get it done fast.

The Connectivity Challenge

If you've ever tried to send a text message or post to social media during a big game, you know firsthand how stadium connectivity can go from bad to worse as soon as the crowd files in. What seems like a simple task. getting online in a modern venue. becomes an exercise in frustration. Now imagine trying to stream high-quality audio or video in that same environment. It's a nightmare scenario that has vexed engineers, broadcasters, and fans for years.

Stadiums present a perfect storm of connectivity challenges. You've got tens of thousands of people packed into a concrete and steel structure, all trying to access the same networks simultaneously. At the 2025 Super Bowl Verizon reported over 38TB of data use and ATT reported 29TB used for a total of over 67TB of data.

Let's take a moment to explore just how much data that really is....

Example 1: Streaming HD Video
- 1 hour of HD video ≈ 3 GB
- 67 TB = 22,333 hours of HD streaming
- That's nearly 2.5 years of non-stop video playback. enough to binge the entire NFL season hundreds of times over

Example 2: Uploading Smartphone Photos
- 1 high-res photo ≈ 5 MB
- 67 TB = 13.4 million photos
- That's like every fan in a 70,000-seat stadium uploading 190+ photos each

Example 3: Streaming Audio (Spotify, podcasts, etc.)
- 1 hour of audio ≈ 100 MB
- 67 TB = 670,000 hours of streaming
- That's 76 years of continuous playtime. or the equivalent of playing "Eye of the Tiger" on loop more than 5 million times

Example 4: Downloading Mobile Apps
- Average app size (e.g., Uber, Instagram) ≈ 250 MB
- 67 TB = 268,000 app downloads
- Enough for every fan in a sold-out stadium to install 3–4 major apps each. just during the game

What about old-school Transister Radios?

The traditional fallbacks don't work that well either. AM/FM radio, which fans relied on for decades to get real-time commentary in stadiums, has become increasingly unreliable. Modern stadium designs with their metal superstructures, enclosed concourses, and multiple tiers effectively create Faraday cages that block radio signals in many seating areas. I've walked through newer stadiums with a portable radio and watched the signal drop to static in entire sections.

What about WiFi?

Stadium WiFi, despite massive investments, still buckles under peak loads. Imagine trying to serve high-speed internet to a small city's worth of people, all in the same 10-acre plot of land. Even state-of-the-art WiFi 6 deployments struggle when everyone's trying to upload selfies after a touchdown. At SoFi Stadium in Los Angeles, they installed over 2,500 WiFi access points. one for roughly every 28 seats. and still face connectivity issues during major events.

What about Cellular?

Then there's the cellular option. Carriers have improved in-stadium coverage with dedicated DAS (Distributed Antenna Systems), but they're fighting a losing battle against spectator behavior. Every year, fans consume more data, stream more video, and expect faster connections. The infrastructure simply can't keep pace with demand. Add in network overhead, and even 5G connections start to falter when 70,000 people are all generating traffic.

The Latency Ladder

The following chart summarizes latency ranges by category, along with representative companies and technologies. As you'll see, sub-second latency isn't just a technical brag, it defines the difference between being live and being late.

5. Why Audio Is the Right Medium

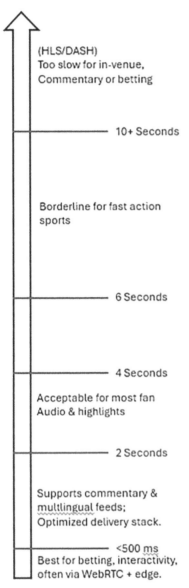

(HLS/DASH)
Too slow for in-venue,
Commentary or betting

— 10+ Seconds

Borderline for fast action
sports

— 6 Seconds

— 4 Seconds

Acceptable for most fan
Audio & highlights

— 2 Seconds

Supports commentary &
multilingual feeds;
Optimized delivery stack.

— <500 ms
Best for betting, interactivity,
often via WebRTC + edge.

There's a reason this chapter isn't called "The Future of In-Stadium Video." Video might be flashy, but audio is the workhorse, and in the stadium environment, it's the far more practical, scalable, and impactful choice.

Start with bandwidth. Delivering a real-time video stream to thousands of fans sitting shoulder to shoulder under concrete and steel is a logistical nightmare. Audio, on the other hand, is lightweight. You can deliver a crisp, clear commentary feed in a fraction of the data footprint. It doesn't require 5G or perfect Wi-Fi. Even in a noisy RF environment, audio can punch through.

It's also personal. Most people already have the gear they need, their phone and a pair of earbuds. That makes adoption frictionless. No special hardware, no kiosks, no sanitation stations. Just a tap and they're in.

More importantly, audio enhances the live experience without competing with it. You can look at the field, feel the crowd, soak in the atmosphere, and still get the expert breakdown in your ear. Video demands attention. Audio gives context without distraction.

From a content standpoint, audio is flexible. You can offer alternate feeds: home team, away team, neutral analysts, different languages. You can add live betting commentary, player stats, or even mic'd-up coaches. You can monetize it with sponsor reads, pre-rolls, or premium tiers. And you can turn it off just as easily, giving the fan full control.

It also travels better. From VIP suites to upper-deck seats, audio can reach everywhere at the same time, without needing a screen in sight. It equalizes the experience, making the same commentary available to the fan in row one and the one in the last row of the nosebleeds.

Accessibility is another area where real-time audio excels. For visually impaired fans, synchronized play-by-play can turn the stadium from a blur of noise into a fully realized experience. No need for bulky IR headsets, advance registration, or special seats. Just tap and listen. And it's not just good inclusion policy, it's ADA-compliant by default. The same infrastructure also supports multilingual streams, giving international fans the context they need, when they need it, without missing a beat. Whether you're in New York or Newcastle, the ability to hear the game in your own language, in sync with the action, isn't just a bonus. It's the baseline.

Real-time audio also creates new opportunities through multilingual commentary. In an increasingly global sports market, venues are filled with fans who speak a range of native languages, from international tourists to second- and third-generation locals. Giving those fans access to the game in their own language isn't just a courtesy. It's a way to foster loyalty, drive deeper emotional engagement, and convert one-time attendees into lifelong fans. It also opens up targeted sponsorship and content opportunities, enabling brands to reach fans more effectively with culturally relevant messaging. In short, language isn't just inclusion. It's strategy.

And because language is personal, audio lets teams connect on a global scale. With real-time multilingual commentary, a visitor from Tokyo or São Paulo can experience the game not as a spectacle, but as a story. That's more than just good UX, it's how you turn tourists into fans.

Put simply, audio is the path of least resistance, technically, operationally, and experientially. It offers the highest impact for the lowest lift.

If you're trying to fix the fan experience without blowing up your network, retraining your staff, or spending millions on hardware, start with audio.

6. Why This Hasn't Been Solved Yet

If real-time audio is such a clear win, for fans, teams, venues, sponsors, and accessibility, then why isn't it already everywhere?

The short answer: fragmentation. The long answer: politics, priorities, and legacy systems that were never built for this use case.

Let's start with the venue. Most stadiums are a patchwork of ownership models, outdated wiring, and multiple layers of stakeholders. A team may not control the Wi-Fi network. The league may own the rights to the in-game audio. The venue operator might outsource digital signage and concessions to a third-party vendor with its own tech stack. Adding something "simple" like real-time audio often means negotiating with five different parties before you ever write a line of code.

Then there's the broadcast rights issue. Many leagues operate under complex agreements that tightly control who can distribute commentary, in what form, and to whom. Even if a team wants to offer in-venue audio through its app, it may be blocked by regional or national licensing restrictions. This is especially true when radio affiliates or over-the-top platforms have exclusive deals in place. So the problem isn't always technical, it's legal.

Another issue is that real-time audio has traditionally fallen into a gray area. It's not glamorous like AR replays or drone camera shots. It doesn't drive ticket sales directly, so it often gets skipped in the budgeting cycle. And unlike security, concessions, or HVAC, it doesn't fall squarely into any one department's responsibility. As a result, it becomes an orphaned initiative, everyone agrees it's a good idea, but no one owns it.

There's also fear of change. Stadium IT teams are already overwhelmed trying to keep POS systems running, fan Wi-Fi stable, and digital displays synced up. They're also battling a sea of technical noise, tens of thousands of personal devices, unauthorized mobile hotspots, rogue Wi-Fi extenders, and Bluetooth peripherals all contribute to radiofrequency congestion that can cripple even the best-laid networks.

In some stadiums, the interference from fans' phones alone is enough to disrupt mesh systems, delay data syncs, and render guest Wi-Fi nearly unusable. Adding another system, even a lightweight one like real-time audio, can be met with either skepticism, or, 'we already did that.'

And that skepticism only deepens when they've been burned before by tech that promised more than it delivered, AR games that crashed, mobile tickets that failed under load, or fan apps that drained batteries and clogged help desks. For IT departments operating under intense pressure on game days, stability almost always beats innovation.

Finally, many attempts at solving this have required too much from the fan. Download an app. Create an account. Enter a code. Accept permissions. Sync with your ticket. The more steps you add, the more people drop off. Any friction at all can kill adoption.

The irony is that none of these problems are insurmountable. But they require a solution that understands the landscape, one that slots into existing workflows, doesn't break rights agreements, and requires zero behavioral change from fans.

The technology is ready. The fan appetite is real. The final hurdle is organizational inertia and structural barriers within the sports ecosystem itself.

7. Who's Going to Win This?

If we've learned anything from the first wave of internet broadcasting, it's that the winners aren't always the flashiest, the biggest, or even the best-funded. The winners are the ones who make it easy. That was the case with broadcast.com. It seemed like every day we saw press releases from some east or west coast startup with lots of investor money and flashy demos and flashier promises. Meanwhile we soldiered on in Dallas, making it happen.

In the race to own real-time in-stadium audio, the field is wide open. There are infrastructure companies, app developers, broadcast giants, and upstart platforms all circling the same opportunity from different angles. But they're not all solving the same problem. Some are focused on latency. Others are optimizing for scale, rights management, interactivity, or analytics. A few are just hoping to ride someone else's rails.

The solution that ultimately prevails will be the one that delivers the right balance, fast enough to feel live, simple enough to require no training, scalable enough to work across dozens of venues, and smart enough to integrate into the larger app ecosystem without disrupting it.

It won't be the one with the most patents, the slickest demo, or the flashiest booth at NAB. It'll be the one fans use without thinking. The one that feels inevitable once you try it. The one that works the first time, every time, regardless of whether you're in row 5 or row 505.

And just as Broadcast.com found its footing by solving a real-world problem, making radio listenable over the internet, the next leader will emerge by solving the real problem fans face in stadiums today: silence.

At Broadcast.com, my fundamental rule was that we treated every event like a broadcaster would: failure simply wasn't an option. Unlike traditional IT approaches where occasional downtime might be acceptable, we understood that in broadcasting, when the signal drops, you've failed your audience completely. We built our entire methodology around three core principles: flexibility to adapt to changing conditions, redundancy to ensure continuous service, and speed to deliver content when and where it mattered.

Even today, I see companies approaching stadium audio as if it were merely an IT problem to solve rather than a broadcasting challenge. They focus on the technology without understanding the unforgiving nature of live events. The next market leader will be the one who recognizes, as we did at Broadcast.com, that this is fundamentally about reliable content delivery that cannot fail, addressing the real and immediate problem fans face in stadiums today: silence.

There's still a frontier to be claimed. It's not video. It's not VR. It's not flying cars or holograms. It's something simpler, more human, and more immediate.

It's giving voice to the game, in the moment, where it matters most.

8. A Note About SportsBug

After spending this chapter examining the in-stadium audio challenge from an industry-wide lens, I would be remiss if I didn't acknowledge the work I've personally been doing to help solve it.

I am the CEO and a cofounder of SportsBug, a company that traces its roots back to the 1980s when John Roe first

launched the original SportsBug headset. Back then, it was an AM radio earpiece you bought at the stadium to hear the live play-by-play, later evolving into FM. It gave fans what they craved: real-time context to match the real-time action. But as stadiums got bigger, denser, and more hostile to radio signals, that model faded. The technology couldn't keep up with the concrete, steel, and cellular congestion of modern venues.

What we've built today is a reinvention, not a revival. The new SportsBug is a streaming-first, mobile-native, zero-latency platform that delivers real-time play-by-play audio directly to a fan's smartphone, with no additional infrastructure, no special hardware, and no noticeable delay.

Unlike traditional streaming that can lag by 30 seconds to two full minutes or more, or AM/FM feeds that are blocked by structural interference, SportsBug leverages proprietary technology and the full power of 4G and 5G networks to deliver audio with sub-one-second latency. Tests across stadiums in the U.S. and abroad have consistently shown latency under 400 milliseconds. In side-by-side comparisons with league apps and station streams, the difference is not just measurable, it's game-changing, in some cases, several minutes.

But it's not just about tech for tech's sake. What makes SportsBug different is that it solves the problem without creating new ones. There is no cost to the team, league, or venue. The monetization comes from sponsor packages and branded headphone sales, not licensing fees or installation budgets. It's a BYOD model, bring your own device, press play, and you're in. We've even partnered with Learfield, the largest college sports media group in the country, to integrate into their Varsity Network App, bringing our zero-delay experience to fans in stadiums nationwide.

If you're looking for a case study in how to do real-time audio right, I believe SportsBug is that proof of concept.

The industry data confirms we're addressing a real need. With 82% of sports fans using their mobile phones during live events and 72% preferring real-time audio updates, there's clear demand for what we're offering. Our case studies demonstrate that fans are frustrated with existing solutions, with official apps routinely experiencing delays of half a minute to several minutes.

Our vision extends beyond just solving today's problems. We're building a comprehensive platform that will evolve with changing technologies and fan expectations. From data analytics that provide valuable insights to teams and sponsors, to content aggregation that expands our reach beyond the stadium, we're positioning SportsBug as a central player in the future of sports media.

This isn't just a market ripe for disruption. It's a problem whose time has come to be solved. And after decades of watching this space from both inside and outside, I decided it was time to stop waiting for someone else to fix it.

Note: In the appendix, I've included a reference guide to the companies currently operating in this space, including both service and software providers. This resource provides detailed profiles and competitive analysis that will be valuable for industry professionals, researchers, and anyone seeking a deeper understanding of the technical landscape. I encourage those with professional or academic interest in this field to review this material as a starting point for further investigation.

Conclusion: Sync or Sink – The Next Chapter of the Streaming Wars

What started as a scramble to get video online has become a battle for something deeper, the last mile of experience. Chapter 15 hasn't just been about latency charts or encoding tricks. It's been about how we close the gap between what's happening and what's being felt. That gap, that delay, has always been the real enemy. And in many ways, this is the logical endgame of everything this book has explored: control, access, timing, and presence.

In the '90s, we just wanted the stream to play. We wired radios with borrowed gear and prayed RealAudio wouldn't crash. Multilingual overlays and browser-based audio in under 300 milliseconds weren't even on the horizon. But the instinct that drove us then, the need for immediacy, is still what drives innovation now. Back then, buffering meant you were pushing boundaries. Today, it means you've already lost.

The companies in this chapter are chasing that same instinct. Some are building WebRTC meshes at continental scale. Others are making it possible for your father to hear the game in his own language without touching a settings menu. But they're all solving the same problem: how to keep up with the moment.

We're in an era where fans expect personalization without friction, immersion without delay, and commentary that feels more like a conversation than a broadcast. It's no longer enough to stream the game. You have to stream the feeling of being there. And that means solving for sync, voice, context, and presence, on devices that were never built for any of this.

Real-time streaming isn't just a technical milestone. It's a competitive differentiator. It's the difference between watching a game and living it.

And that brings us full circle. Because real-time audio isn't just the next feature on a roadmap. It's the newest battleground in a streaming war that never really ended. A war about who owns the rails, who controls the rights, and who gets to define the fan experience.

We opened this book with a spinning buffer icon and a pixelated screen. We're ending it in a stadium full of fans, some in the seats, some half a world away, all hearing the same play, at the same time, in their own language, on their own terms.

That's not just progress. That's presence. And it didn't happen by accident. It happened because people built it, through trial, collapse, reinvention, and persistence.

So, if the early streaming wars were fought over file formats, codecs, and bandwidth, the next chapter is simpler and more human.

Can you keep up with the moment?
Because if you can't, someone else will.

16: Beyond Broadcast.com

The Next Frontier in AI-Powered, Real-Time Sports Streaming

When we began this journey, the idea of watching a game on the internet was borderline science fiction, or at least the stuff of late-night demos that barely held together. Broadcast.com was a breakthrough not because it was perfect, but because it proved that fans wanted more than just access to sports. They wanted agency. They wanted immediacy. They wanted what *they* wanted, *when* they wanted it, not when the TV Guide said they could.

Now, after decades of evolution, after the codec wars, the rise and fall of flash, the slow death of buffering, we find ourselves at the brink of something different. The next wave isn't about getting video online. That part's done. The next wave is about *closing the loop*. Making what you see, what you hear, and what you feel line up with what's actually happening, in real time.

And not just for one type of fan, in one location, on one network. For *everyone*.

Streaming Isn't the Product Anymore. The Experience Is.

We used to measure progress in terms of pixels and framerates. But what really matters now isn't how sharp the video is. It's how close it feels. Does it respond to you? Does it adapt? Can it anticipate what you want before you even reach for the remote?

We're leaving behind the era of one-size-fits-all broadcasts and stepping into a landscape defined by personalization, synchronization, and immersion. In this environment, whoever delivers presence, the feeling of being there, connected to the action as it unfolds, wins.

That means real-time audio. It means predictive AI. It means overlays, translation, interaction, and participation. And critically, it means zero tolerance for lag.

AI: Not Just Smarter: Faster

Artificial intelligence won't replace broadcasters. But it will replace *delay*. We're already seeing systems that:

- Automatically generate highlights based on a viewer's favorite player.
  - ○ **WSC Sports:** AI-driven platform used by the NBA, Bundesliga, and MLS to automatically generate and personalize highlight reels. https://wsc-sports.com
  - ○ **IBM Watson Media:** *(now part of IBM Video Streaming)* IBM used to market this under the "Watson Media" brand, but it's now consolidated under **IBM Video Streaming** and **Watsonx AI** initiatives. Their video enrichment tools, including automated highlights, closed captioning, and metadata tagging, are still being used in broadcast and sports... https://www.ibm.com/products/video-streaming For AI-driven automation within video workflows (e.g., highlight detection, sentiment analysis), they now emphasize https://www.ibm.com/watsonx
  - ○ **Pixellot:** Offers automated game capture and highlight generation for youth and semi-pro sports. https://www.pixellot.tv
- Offer translated commentary in multiple languages, rendered in real time.

- o **Microsoft Azure AI Speech Translation:** Real-time translation of spoken language; used in sports events and conferences. https://azure.microsoft.com/en-us/products/cognitive-services/speech-translation
- o **Google Cloud Media Translation API:** Real-time streaming translation that can be embedded in live commentary platforms. https://cloud.google.com/media-translation
- o **Kiswe Studio:** Used by the NBA and global concerts to deliver alternate language commentary tracks, often paired with real-time translation tools. https://www.kiswe.com/studio
- Respond to natural language questions mid-game ("What's that formation?" or "When did she last score from that angle?").
  - o **Stats Perform's Opta GPT (under development)** Leverages Opta's data and LLMs (like ChatGPT) to allow fans to ask live, natural-language questions about players, formations, history, etc. https://www.statsperform.com
  - o **Sportradar & Genius Sports** (exploring similar capabilities): Not direct URLs for consumer use yet, but both are integrating AI-driven analytics for broadcasters and betting platforms.
  - o **Amazon Alexa Sports & Google Assistant:** These are more generalized, but both are slowly expanding real-time sports query capabilities via voice assistants.

The core advantage isn't novelty, it's *speed*. AI can shrink the gap between what's happening and what's understood. And in a live sports context, understanding *now* is everything.

AR: The Layer That Changes the Game

Augmented reality is no longer a gimmick. It's becoming the interface layer between live sports and the digital enhancements fans want.

- **In-Stadium Experiences:** Real-Time Player Stats and Heatmaps: Fans can use their smartphones to access live data overlays, such as player statistics and heatmaps, directly on the field during matches. For instance, some football clubs are testing heads-up displays that provide tactical insights and injury updates https://www.technolynx.com/post/augmented-reality-in-football-a-new-era-of-fan-engagement

- **At-Home Viewing:** Interactive TV Overlays: Broadcasts can incorporate AR to display advanced metrics, betting odds, and historical comparisons on viewers' screens. This integration offers a more immersive and informative experience, allowing fans to engage deeply with the game from their living rooms.
- **Group Settings in Public Spaces:** Shared AR Experiences: In venues like sports bars, AR can transform group viewing by providing interactive features. Fans can participate in multiplayer AR games, access synchronized live stats, and enjoy collective virtual experiences that enhance social interaction and engagement during live events. https://aroundar.com/

AR won't replace video. It will contextualize it. It's the difference between watching the game and understanding it as you watch.

Real-Time Audio: The Fastest Path to Presence

Among all these technologies, real-time audio remains the dark horse, lightweight, scalable, and incredibly effective.
- It requires less bandwidth than video.
- It works across venues, devices, and languages.
- And when done right, it brings fans into the moment faster than anything else.

We've seen that in-venue, real-time commentary fills the silence left by muffled or simply drowned-out PA systems and ambient chaos. For remote fans, synchronized audio prevents spoilers from crowd noise or push alerts. And for global audiences, real-time multilingual feeds remove a long-standing barrier to full participation.

If streaming video is the canvas, real-time audio is the brushstroke that gives it feeling.

What This Means for the Industry

For Investors:

This isn't speculative anymore. Real-time engagement platforms are forming the core of next-generation media companies, with monetization pathways that include sponsorship, betting, gamified interactions, and dynamic advertising. AI and AR aren't optional add-ons, they're *infrastructure*.

For Analysts:

The shift is happening in the metrics. Viewership is no longer the north star. *Engagement density*, fan dwell time, language toggles, and interactive moments per minute are becoming important indicators of success. The ability to track, react to, and learn from those signals will define the winners.

For Broadcasters, Leagues, and Teams:

The competitive line has moved. Fans don't compare your stream to the one from five years ago, they compare it to TikTok, Twitch, and the last push notification they got from a fantasy app. You're not just competing with other games. You're competing with everything else their attention could land on.

If you don't own your experience layer, someone else will build it around you, and monetize the audience you thought you had.

What's Next: A Stack Built for Presence

We're not talking about standalone features anymore. We're talking about a stack:
- Low-latency delivery
- AI-powered curation

- Multilingual, personalized commentary
- Adaptive overlays and second-screen interactions
- Context-aware fan tools and predictive services

Each layer strengthens the next. Each innovation shortens the distance between the game and the fan.

And when those layers work together, they create something we've been chasing since the dial-up days: a sense of now. Not "almost live." Not "good enough for mobile." Just **_now_**.

Conclusion: The Stream Becomes the Stadium

We started this book by tracing the early architecture of streaming, how engineers, entrepreneurs, and misfits bootstrapped a media revolution. Along the way, we saw how control shifted from studios to servers, from networks to nodes, from broadcasters to devices.

But the final shift, the one we're witnessing now, is from passive access to active experience.

The stream isn't just a way to watch the game. It's becoming the *stadium itself*. The place where fans gather, engage, react, and belong.

That's the new frontier. It's not about replacing the old way. It's about fulfilling the promise we glimpsed years ago: a world where every fan, everywhere, can be part of the game, not just a viewer, but a participant.

And this time, we're not buffering.

AFTERWARD

The Signal That Got Through

If you've made it this far, thank you. Seriously. Thank you for taking the ride. For stepping back into the chaos, the breakthroughs, the backrooms, and the buffer wheels that built the strange, brilliant world we now call streaming media.

This book wasn't meant to be a victory lap. It's a field guide. A chronicle. A flashlight pointed at the things we got right, and the things we still haven't figured out. And at the heart of it all is something simple: the moment.

Because that's what we've all been chasing, from the earliest days of hacked-together codecs and second-story office installations to the real-time overlays and predictive engines of today. The moment when a fan connects. When the action and the emotion line up. When the stream stops being just a feed, and becomes the experience.

That moment isn't just a technical target. It's why we do this.

And look, I'll be honest, there's no roadmap for what comes next. We're building it as we go. We're mixing AI with human instinct. We're piping multilingual audio into stadiums. We're giving fans control over how they watch, how they hear, and maybe soon, how they shape the game itself.

That's the fun part. The magic, if I can say that without rolling my eyes. Building things no one's built before. Solving problems in ways that would've gotten you laughed out of a conference room ten years ago. Or maybe twenty.

We're not just innovating. We're inventing. And that's a rare thing.

If you're in this space, whether you're building, investing, advising, or just trying to understand it, I hope this book helped you see the bigger picture. Not just how far we've come, but how much is left to do. And how much fun it can be to do it alongside people who still care about getting it right.

And if you liked this journey, you might be surprised to hear I write fiction too. At the time of this writing, I'm working on Book 5 of my Accipiter War science fiction series, which I co-author with my son, Blake Seaman. It's a very different kind of world, alien empires, warships, human survival, but at its core, it's about the same thing:

---

signal through the noise.

Staying connected. Fighting for the future you believe in.

Sound familiar?

Thanks again for reading. Stay curious. Stay connected. And remember, whatever comes next, build like it matters.

■ Patrick Seaman

• 302 •

Appendix A: Real-Time Audio & Streaming: Market Approaches and Technology Landscape

Introduction and Disclaimer

This appendix provides a high-level overview of the current landscape for real-time audio and streaming technologies within the live sports and event industries. It compiles information from publicly available sources including company websites, public presentations, press releases, case studies, and trade media. No confidential or proprietary information was used or inferred. This is not an endorsement, ranking, or formal evaluation of any company or product.

The technologies and players in this space are evolving rapidly. The purpose of this section is to help contextualize the range of approaches and infrastructure models being used, and to illustrate the diversity of strategic thinking shaping the next generation of in-venue audio and streaming experiences.

Technology Clusters Overview

Rather than listing companies individually, it is more instructive to group them by common approach or architectural philosophy. Below are four major functional clusters that define the current market based on my personal experience and observations. This list includes examples of current and historical entrants that I've directly encountered or researched across the spectrum. It's important to note that this isn't intended to be a globally comprehensive or exhaustive inventory of all solutions. Any exclusions from this list are not intentional. Not all entities mentioned are still active, and their inclusion here reflects their relevance to the category's evolution as I've witnessed it throughout my career in this space.

1. In-Venue Experience Platforms

These solutions are optimized for fans inside a stadium, typically using private Wi-Fi, cellular, RF, or hybrid systems to deliver synchronized, sub-second audio and visual enhancements.

- **Mixhalo**: Wi-Fi-based low-latency audio for live sports and music venues.

- **DVOX**: RF transmitters + app-based sync for motorsports and tennis zones.

- **ProWire**: Browser-based low-latency commentary with QR-code onboarding.

- **SportsBug**: Ultra-low-latency app-based audio with geofenced access.

- **Cue Audio**: Data-over-audio tones for synchronized fan games and light shows.

- **D2N**: RF + IP hybrid systems for venues and events, especially in Australia.

- **VOGO**: Multicam replays, slow motion, and referee audio overlays for fans.

2. Streaming Infrastructure & Developer Tools

Platforms offering back-end tools, SDKs, or APIs to power low-latency audio/video experiences across apps and custom platforms.

- **Phenix**: Scalable WebRTC mesh for real-time betting and live sports.

- **Agora**: SDK-based real-time engagement with global sub-second latency.

- **Dolby.io**: Millicast-powered WebRTC for immersive, high-fidelity media.

- **Red5 Pro**: Autoscaling edge mesh built for WebRTC and global reach.

- **Wowza**: Long-established, protocol-agnostic media server with hybrid latency support.

- **Ateme**: Encoding and distribution infrastructure for broadcast-grade delivery.

- **Harmonic**: SaaS-based media processing and encoding for OTT and edge delivery.

- **Zixi**: Broadcast-grade, low-error video delivery infrastructure using proprietary protocols.

3. Experimental or Retired Models

Historical or now-defunct efforts that influenced second-screen and real-time innovation.

- **Tunity**: Used audio fingerprinting to sync TV streams in public venues.

- **League-Specific Apps (e.g., NBA, MLB)**: Early in-app radio with variable latency.

- **Verizon Media / Edgecast**: Edge CDN with low-latency video workflows, now absorbed into Yahoo.

SWOT Analysis Overview

The following SWOT matrix is intended to summarize the broader dynamics shaping the real-time in-venue audio landscape. Rather than evaluating any one company or product, it highlights the recurring themes, structural advantages, and shared obstacles that are emerging across the ecosystem.

This high-level view is designed to help teams, leagues, investors, and technology partners understand where the most consistent strengths lie, what challenges are still unresolved, and where opportunity may exist, not just for technical innovation, but for commercial traction and long-term viability.

STRENGTHS	WEAKNESSES
Real-time audio solves a widespread fan pain point	Infrastructure limitations in older venues
Low barrier to adoption via mobile-first delivery	Highly fragmented rights and broadcast policies
Enhances accessibility, inclusivity, and multi-language support	Fan education and onboarding still evolving
OPPORTUNITIES	THREATS
Global expansion through multilingual commentary	Conflicts with existing broadcast or IP rights
Sponsor integration and branded hardware	Technical interference and congested RF/Wi-Fi environments
ADA and accessibility compliance for public funding	Competing initiatives that silo fan attention inside apps

Technical Tradeoffs Matrix

Every company in the real-time streaming space makes tradeoffs, between latency and scale, between simplicity and control, between friction and flexibility. This matrix outlines the most common technical decision points faced by vendors and integrators working to deliver sub-second audio and video experiences.

Rather than prescribing a single "best" approach, the goal here is to illuminate how different solutions prioritize latency, scalability, deployment complexity, and device compatibility. Understanding these tradeoffs can help decision-makers align their technical strategy with fan expectations, venue realities, and operational constraints.

Latency vs. Scalability

- WebRTC: Ultra-low latency (<500ms), but complex at scale

- Wi-Fi Audio: Extremely low latency, depends on local network quality

- Cloud/CDN: Easier to deploy, introduces buffering and lag

Delivery Mechanism

- App-based: Full control, better UX, but requires install

- Browser-based: Lower friction, but limits interactivity

- RF Hybrid: Requires setup, but extremely reliable once deployed

Edge Strategies

- Some solutions pre-cache or buffer content at the device level

- Others use peer distribution or local repeaters to reduce load

Past Concepts and Lessons Learned

Several early or defunct entrants provided proof-of-concept for real-time audio's appeal, even if they didn't endure.

- **Tunity** validated passive audio consumption in public venues, but ran into IP and monetization friction

- **League-run apps** proved inconsistent when network conditions varied or when apps prioritized video features over latency sync

This overview is not exhaustive but offers a representative snapshot of the technological landscape and market experimentation surrounding real-time in-venue streaming and audio. For those evaluating or building in this space, understanding the patterns is more important than tracking the players.

Appendix B: Accessibility and Language – Inclusion That Scales

While much of the conversation around real-time audio focuses on latency, scale, and monetization, its most transformative potential may lie in access and inclusion. Stadiums and arenas are public spaces, cultural landmarks, and, increasingly, symbols of modern infrastructure. As expectations shift from passive attendance to personalized, tech-enhanced participation, accessibility can no longer be treated as a compliance box to check. It must become a design principle.

This appendix explores how real-time audio meets that moment. It supports fans with sensory impairments, welcomes international audiences, and helps teams and venues stay ahead of tightening global regulations. But more than that, it unlocks a more human experience, one where everyone hears the moment as it happens, and in a voice they understand.

Access Isn't Optional Anymore

Real-time audio doesn't just improve the fan experience. It meets a growing global demand for equity in live events, and it does so with a combination of elegance, scalability, and compliance-ready design.

From ADA requirements in the United States to the upcoming European Accessibility Act, accessibility is no longer a "nice to have", it's baked into the expectations of modern venues, broadcasters, and fans.

Inclusion Without Friction

For fans with visual impairments, synchronized audio descriptions transform the experience from chaotic to coherent. Instead of watching in silence or relying on lagging radio feeds, real-time commentary puts them on equal footing with everyone else, without requiring special hardware, dedicated seats, or advance notice. BYOD (bring your own device) delivery avoids hygiene concerns, maintenance costs, and staff training, all while offering scale that traditional systems can't match.

Multilingual fans face a different kind of exclusion: linguistic. Stadiums may host global audiences, but they often deliver only one or two language options, if any. The result? Tourists and immigrant fans, who tend to spend more on merchandise and upgrades, are left following along without context. Real-time multilingual audio lets them hear the game in their own language, in sync with the moment.

How Real-Time Audio Delivers Compliance

While the experiential benefits of real-time audio are obvious to fans, its legal and operational implications are just as significant. Around the world, stadiums, teams, and broadcasters face increasing pressure, and in many cases, formal requirements, to deliver accessible experiences for all attendees. Real-time audio isn't just a tool for engagement, it's a scalable solution that helps meet a patchwork of evolving accessibility laws across jurisdictions. Below is a breakdown of how real-time audio aligns with major compliance frameworks and accessibility mandates.

United States, ADA (Americans with Disabilities Act)

- Titles II and III require "effective communication" for public spaces, including stadiums.

- Real-time audio qualifies as assistive communication, especially when descriptive services are included.

- BYOD systems align with Department of Justice guidance by eliminating need for dedicated personnel or devices.

European Union, EAA (European Accessibility Act, Directive 2019/882)

- Effective mid-2025, requires accessibility for digital and media services.

- Real-time audio provides real-time parity and localization across multilingual EU audiences.

- Additional compliance touchpoints: UK Equality Act, French Law No. 2005-102, Germany's BITV 2.0.

Global Frameworks

- Canada: Accessible Canada Act (barrier-free services by 2040)

- Australia: Disability Discrimination Act 1992

- United Nations Convention on the Rights of Persons with Disabilities (CRPD), ratified by 185+ nations

  - Article 30: Equal access to culture and sport

  - Article 9: Information accessibility

These frameworks create not just obligations, but opportunities, to lead on ESG, to qualify for funding and public-private partnerships, and to demonstrate meaningful inclusion to fans and sponsors alike.

Multilingual Commentary: From Broadcasts to Belonging

Language is emotional. It's personal. When fans can follow every play, every nuance, every moment in their own language, the game stops being a broadcast, and starts being theirs.

Real-time multilingual commentary unlocks:

- **Global reach**: Expands the audience across time zones and continents

- **Sponsorship precision**: Enables region-specific ads and messaging

- **OTT and direct-to-consumer differentiation**: Adds a layer of choice that younger global audiences expect

- **Emotional connection**: Strengthens loyalty, particularly among underserved or first-time attendees

For venues, it's a way to enhance the live experience. For broadcasters and leagues, it's a way to grow. For fans, it's simply a better, more human experience.

Multilingual Sports Commentary: Unlocking Global Engagement

Live sports is one of the few remaining global media experiences, but much of the in-stadium experience is still delivered in a single language, often English. For many international fans, that means attending a game in person can be more spectacle than story. Real-time multilingual commentary bridges that gap.

As travel and team fandom become increasingly globalized, so too does the expectation of inclusion. A Japanese tourist at a Lakers game, a Brazilian supporter at the Champions League final, these fans are deeply invested, often more so than local attendees. Yet they're forced to consume the game without context, commentary, or cultural cues.

Real-time audio changes that. With a simple tap, fans can access the feed in their native language, in sync with the action in front of them. This is more than translation. It's localization, commentary that reflects not just language, but tone, emotion, and understanding.

It's also a monetization lever. Brands can target specific linguistic feeds with culturally relevant ads. Broadcasters can serve international audiences more effectively. Teams can expand their CRM efforts with segmented data based on language and regional behavior.

Multilingual audio is where accessibility meets globalization. And in a market where the next billion sports fans may not speak your language, this isn't an optional feature, it's a growth strategy.

Summary: Accessibility is a Strategic Mandate

Real-time audio is more than a technical solution, it is a strategic enabler that addresses multiple dimensions of exclusion. For fans with visual impairments, it provides the missing context that transforms a disjointed experience into one that is immersive and equitable. For international visitors, it offers the comfort and clarity of their native language. And for all fans, it ensures that being physically present at a game no longer means sacrificing understanding or emotional connection.

By supporting a wide spectrum of needs through a single, scalable tool, real-time audio delivers on the promise of universal design. It eliminates the need for special seating, advance requests, or proprietary hardware. Fans simply bring their own device, press play, and access the same quality of experience as everyone else, no exceptions.

This is not just a user benefit. It is a compliance strategy that future-proofs stadiums, teams, and leagues.

Real-time audio bridges multiple forms of exclusion, sensory, cultural, linguistic, with a single, scalable tool.

It aligns with:

- ADA (U.S.)

- EAA (EU, 2025)

- CRPD and national equivalents (Worldwide)

And it delivers more than compliance. It delivers connection. Which, for sports, is everything.

Appendix C: References

Chapter 1 References:

Regulating Netflix: Cross-National Momentum *Discusses the global regulatory challenges faced by streaming platforms, focusing on local content requirements.* Source: PolicyReview.info, https://policyreview.info/articles/news/regulating-netflix-cross-national-momentum/1706

Policy Issues on the Streaming Video Horizon *Explores taxation, content regulation, and platform dynamics.* Source: Digital Content Next , https://digitalcontentnext.org/blog/2022/10/27/policy-issues-on-the-streaming-video-horizon/

How Web3 is Redefining Content Creation *Explains how decentralized platforms and blockchain are changing monetization and distribution.* Source: TDeFi , https://tde.fi/founder-resource/blogs/blockchain-networks/how-web3-is-redefining-content-creation/

Cable News Fact Sheet *Highlights declining viewership of traditional cable TV in favor of streaming.* Source: Pew Research , https://www.pewresearch.org/journalism/fact-sheet/cable-news/

CNN Faces Unprecedented Crisis After Historic Ratings Collapse *Analyzes CNN's ratings challenges and digital transformation strategy.* Source: EconoTimes , https://www.econotimes.com/CNN-Faces-Unprecedented-Crisis-After-Historic-Ratings-Collapse-Only-61000-Viewers-in-Key-Demo-1694127

CNN Invests in Digital Transformation *Details CNN's streaming platform investments to align with shifting habits.* Source: Financial Times , https://www.ft.com/content/e1368295-b24e-4295-8ff8-7015714286fe

Traditional TV and Streaming Viewers in the U.S. *Provides updated viewership trends comparing streaming and broadcast.* Source: Statista , https://www.statista.com/statistics/1385919/traditional-tv-streaming-viewers-us/

Streaming Stats and Trends *Includes habits, engagement, and ad-supported models across streaming.* Source: Forbes , https://www.forbes.com/home-improvement/internet/streaming-stats/

Streaming Trends and Stats *A marketer's view on the rise of streaming platforms and audience shifts.* Source: GWI , https://blog.gwi.com/marketing/streaming-trends-and-stats/

YouTube's Ghost Town Problem *Reveals that billions of YouTube videos go unwatched, illustrating platform saturation.* Source: TechSpot , https://www.techspot.com/news/106791-youtube-numbers-uncovering-youtube-ghost-town-billions-unwatched.html

Chapter 2 References:

Founding of YouTube YouTube was founded in 2005 by Steve Chen, Chad Hurley, and Jawed Karim, revolutionizing content creation and consumption. https://en.wikipedia.org/wiki/YouTube

First YouTube Video: Me at the Zoo The first-ever video uploaded to YouTube was "Me at the Zoo," featuring co-founder Jawed Karim discussing elephants. https://www.youtube.com/watch?v=jNQXAC9IVRw

Google Acquires YouTube Google announced its acquisition of YouTube for $1.65 billion in stock on October 9, 2006, and finalized the deal on November 13, 2006. https://www.sec.gov/Archives/edgar/data/1288776/000119312506238320/dex991.htm

H.264 Video Compression Technology H.264, also known as Advanced Video Coding (AVC), became the dominant codec for streaming due to its efficiency and quality. https://en.wikipedia.org/wiki/H.264/MPEG-4_AVC

YouTube Partner Program Launched in 2007, the YouTube Partner Program allowed creators to monetize their videos through ad revenue. https://support.google.com/youtube/answer/72902?hl=en

Twitch: Evolution of Live Streaming Twitch, initially a gaming-focused offshoot of Justin.tv, became a major live-streaming platform with over 140 million monthly active users by 2020. https://en.wikipedia.org/wiki/Twitch_(service)

TikTok's Rise to Prominence TikTok, launched in 2016, emphasized short-form, mobile-first videos and gained immense popularity globally. https://en.wikipedia.org/wiki/TikTok

Netflix's Streaming Transition Netflix transitioned from a DVD rental service to streaming in 2007, later becoming a pioneer in original programming with shows like *House of Cards*. https://en.wikipedia.org/wiki/Netflix#History

Netflix's Discontinuation of DVD Rentals Netflix announced it would send out its final red envelope on September 29, 2023, marking an end to 25 years of mailing DVDs to members. https://www.cnn.com/2023/04/20/tech/netflix-dvd-service-shutdown-subscribers/index.html

YouTube's Role in Monetizing User-Generated Content The rise of user-generated content on YouTube introduced new revenue streams for creators through ads and sponsorships. https://en.wikipedia.org/wiki/YouTuber#Monetization

The 'Adpocalypse' of 2017 In 2017, YouTube faced a significant advertising boycott due to concerns over ads appearing alongside extremist content, highlighting the platform's challenges in balancing advertiser interests with creator freedom. https://www.theverge.com/2017/3/24/15053990/google-youtube-advertising-boycott-hate-speech

Broadcast.com and Its Legacy Broadcast.com paved the way for live-streaming media, later acquired by Yahoo! in 1999 for $5.7 billion. https://en.wikipedia.org/wiki/Broadcast.com

House of Cards and Netflix's Original Programming Strategy *House of Cards*, Netflix's first original series, marked the company's entry into high-quality, exclusive content. https://en.wikipedia.org/wiki/House_of_Cards_(American_TV_series)

Cultural Impact of PewDiePie and Other YouTube Creators PewDiePie became a global sensation, epitomizing YouTube's potential for individual creators to achieve massive reach. https://en.wikipedia.org/wiki/PewDiePie

Chapter 2.5 References

Enron & Blockbuster: The Failed Streaming Partnership

- **Enron and Blockbuster's 20-Year Video-on-Demand Deal (2000):**
  Details the ambitious agreement between Enron and Blockbuster to provide movies via TV set-top boxes and PCs.
  https://www.latimes.com/archives/la-xpm-2000-jul-20-fi-55629-story.html

  Termination of the Enron-Blockbuster Partnership:
  Reports on the unexpected end of the exclusive 20-year agreement between Enron and Blockbuster.
  https://www.wsj.com/articles/SB984181374790463655

  Enron's Recognition of $110 Million from the Blockbuster Deal:
  Explains how Enron used mark-to-market accounting to book projected profits from the Blockbuster deal.

https://internationalbanker.com/history-of-financial-crises/the-enron-scandal-2001/

Enron's Collapse and Arthur Andersen's Downfall

- **Overview of the Enron Scandal:**
  Provides a comprehensive summary of the Enron scandal, including the role of Arthur Andersen.
  https://faculty.ksu.edu.sa/sites/default/files/the_enron_scandal_and_the_sarbanes-oxley_act.pdf

  Arthur Andersen's Involvement and Conviction:
  Details the accounting firm's role in the Enron scandal and its subsequent conviction for obstruction of justice.
  https://www.britannica.com/money/Arthur-Andersen

  The Andersen Effect and the Sarbanes-Oxley Act:
  Discusses the increased due diligence practiced by auditors following the Enron scandal and the enactment of the Sarbanes-Oxley Act.
  https://www.investopedia.com/terms/a/anderseneffect.asp

Netflix's Rise and Blockbuster's Missed Opportunity

- **Blockbuster Declines to Purchase Netflix for $50 Million:**
  Recounts how Blockbuster rejected Netflix's offer to sell in 2000.
  https://www.screengeek.net/2024/05/04/blockbuster-turned-down-netflix/

- **Netflix Launches Streaming Service in 2007:**
  Chronicles Netflix's transition from DVD rentals to streaming video.
  https://seat11a.com/blog-the-evolution-of-netflix-from-dvd-rentals-to-global-streaming-leader/

YouTube's Emergence and Impact on Digital Media

- **History and Founding of YouTube:**
  Provides background on YouTube's creation and its founders.
  https://en.wikipedia.org/wiki/YouTube

- **YouTube's Official Launch and Early Growth:**
  Details YouTube's launch and its rapid increase in video views.
  https://www.britannica.com/topic/YouTube

Regulatory Reforms Post-Enron

- **The Sarbanes-Oxley Act and Its Implications:**
  Explains the provisions of the Sarbanes-Oxley Act enacted in response to corporate scandals like Enron.
  https://www.thebalancemoney.com/
  sarbanes-oxley-act-and-the-enron-scandal-393497

- **Congressional Oversight and the Enron Scandal:**
  Discusses how the Enron investigations led to landmark legislation, including the Sarbanes-Oxley Act.
  https://levin-center.org/what-is-oversight/portraits/
  congress-and-the-enron-scandal/

Chapter 3: References

U.S. DVD sales declined by 86% between 2006 and 2019

An overview of the collapse of physical media consumption and the broader shift toward digital ownership models. https://theweek.com/news/media/961866/death-of-dvds-and-decline-of-ownership-in-digital-age

Physical media sales in the U.S. decreased to $754 million in the first half of 2023

Reports the continued drop in physical media sales from $1.05 billion in 2022 to $754 million in just six months. https://www.statsignificant.com/p/the-rise-fall-and-slight-rise-of

In-car CD players are disappearing as top manufacturers phase them out

Automakers are eliminating CD players from new vehicles, further signaling the obsolescence of physical media. https://www.thetimes.co.uk/article/car-cd-players-run-out-of-road-as-all-top-manufacturers-shun-them-2jfx9ngr7

U.S. audiences streamed 21 million years' worth of video in 2023

Nielsen reports a 21% year-over-year increase in video streaming consumption across U.S. audiences. https://www.nielsen.com/insights/2024/streaming-unwrapped-streaming-viewership-goes-to-the-library-in-2023/

Streaming accounted for more than one-third of total TV time in 2022

Streaming gained a 6.3 percentage point increase over 2021, reflecting continued growth in binge-watching culture. https://www.nielsen.com/insights/2022/win-the-streaming-wars-with-bingeable-content/

The decline in physical media raises concerns about media preservation

Streaming content is subject to removal and alteration, threatening long-term access and cultural preservation. https://www.techdigest.tv/2024/07/declining-dvd-sales-threatening-the-future-of-media-preservation.html

Corporate censorship and control over digital content

Explores how private corporations may limit or alter the availability of digital content, raising concerns about transparency and freedom of access. https://en.wikipedia.org/wiki/Corporate_censorship

Concentration of media ownership

Describes how media consolidation among a few corporations can limit diversity and editorial independence. https://en.wikipedia.org/wiki/Concentration_of_media_ownership

Media democracy and ownership concentration

Analyzes the political and cultural consequences of concentrated media ownership and its impact on democratic discourse. https://en.wikipedia.org/wiki/Media_democracy

The 2000s saw a decline in physical music formats

CD sales peaked in 2004 before rapidly declining with the rise of digital downloads and streaming. https://en.wikipedia.org/wiki/2000s_in_the_music_industry

The introduction of HD DVD and Blu-ray in 2006 marked the beginning of the end for standard DVDs Streaming surpassed physical media sales by 2017, signaling a permanent shift in media consumption. https://en.wikipedia.org/wiki/DVD

Vinyl records experienced a 17% revenue growth in 2022

Vinyl sales reached $1.2 billion in 2022, continuing a sixteen-year upward trend in physical music nostalgia. https://www.npr.org/2023/03/10/1162568704/vinyl-outsells-cds-first-time-since-1987-records

Younger generations, including Gen Z, are driving vinyl's resurgence

Gen Z and Millennials are embracing vinyl as a collectible, tactile counterpoint to digital streaming.

https://nypost.com/2024/12/03/lifestyle/
vinyl-records-growing-more-popular-than-cds-streaming/

Vinyl remains a niche market, accounting for just 8% of U.S. recorded music revenues in 2022 Despite its popularity, vinyl sales still represent a small slice of the broader recorded music market.
https://www.statista.com/chart/7699/lp-sales-in-the-united-states/

Chapter 4: References

Globalization of Media and Cultural Exchange

Globalization of media: Impacts and implications in a connected world
This article explores the interconnectedness of media systems across national borders, shaping cultural exchange and communication.
https://www.globalmediajournal.com/open-access/globalization-of-media-impacts-and-implications-in-a-connected-world.pdf

Netflix, library analysis, and globalization: Rethinking mass media flows
An academic study examining how global streaming platforms like Netflix challenge traditional media distribution and influence cultural exchange. https://academic.oup.com/joc/article/72/4/511/6605780

Streaming services are bolstering the international entertainment industry
An analysis of how streaming platforms are expanding their global reach and influencing international markets. https://globaledge.msu.edu/blog/post/57226/
streaming-services-are-bolstering-the-international-entertainment-industry

The learning potential of streaming media: Cultural sustainability in a digital world
Discusses how global streaming platforms influence cultural production diversity and the implications for cultural sustainability.
https://www.frontiersin.org/articles/10.3389/fcomm.2023.1084737/

Content Licensing and Legal Challenges

Online entertainment: A new wave of media globalization?
This article considers global online screen entertainment platforms

within ongoing debates about media globalization. https://ijoc.org/index.
php/ijoc/article/download/5725/1832

Is the globalization of content a good thing?

An exploration of the effects of content globalization on technology, the global economy, and pop culture. https://dot.la/content-globalization-2659461163.html

Streaming platforms and their impact on global television culture

Analyzes how services like Netflix, Hulu, and Amazon Prime Video have transformed content consumption, production, and distribution worldwide. https://www.usresistnews.org/2023/11/30/streaming-platforms-and-their-impact-on-global-television-culture/

Content Subscription Future

Factors affecting online streaming subscriptions

A study identifying key drivers behind consumer adoption and retention of streaming services. https://scholarworks.lib.csusb.edu/ciima/vol16/iss1/2

Analysis of the factors influencing streaming video services' business performance

Presents findings from a 2023 international conference examining data and performance metrics in the streaming industry. https://www.scitepress.org/Papers/2023/128086/128086.pdf

TikTok and the live shopping revolution

Explores how TikTok is redefining real-time commerce through livestreaming and integrated shopping features. https://www.thetimes.co.uk/article/tiktok-and-the-live-shopping-revolution-enterprise-network-n6z6f9mb0

NBCU's streaming chief isn't worried about you canceling cable

An interview with Peacock's Matt Strauss on the shifting economics of sports, streaming, and subscriber strategy. https://www.theverge.com/24258301/matt-strauss-peacock-nbc-comcast-olympics-sports-direct-to-consumer-video-streaming

Chapter 5: References

Broadcast.com

Broadcast.com General overview and historical background on the company founded by Mark Cuban and Todd Wagner. https://en.wikipedia.org/wiki/Broadcast.com

Broadcast.com – Britannica entry Encyclopedic summary of the company's founding, IPO, and acquisition by Yahoo!. https://www.britannica.com/topic/Broadcast-com

Broadcast.com Goes Public A D Magazine retrospective on the IPO of Broadcast.com and the media climate of the late 1990s. https://www.dmagazine.com/publications/d-magazine/2010/january/broadcast-com-goes-public/

Mark Cuban Sells Broadcast.com to Yahoo A summary of the Broadcast.com acquisition deal and its implications for Cuban's future. https://www.benzinga.com/general/education/21/04/20446634/this-day-in-market-history-mark-cuban-sells-broadcast-com-to-yahoo

How Todd Wagner and Mark Cuban Sold Broadcast.com to Yahoo for $5.7 Billion A podcast episode from The Learning Leader Show exploring the background and execution of the deal. https://learningleader.com/episode/004-how-todd-wagner-and-mark-cuban-sold-broadcast-com-to-yahoo-for-5-7-billion/

Napster

Napster – Wikipedia Overview of Napster's history, including its launch, legal battles, and transformation into a licensed service. https://en.wikipedia.org/wiki/Napster

Napster – Britannica entry A high-level look at Napster's impact on the music industry and its legal downfall. https://www.britannica.com/topic/Napster

Napster: The File-Sharing Service That Started It All An NPR feature tracing Napster's cultural and technological legacy. https://www.npr.org/2009/12/21/121690908/napster-the-file-sharing-service-that-started-it-all

The Napster Revolution A Guardian article exploring the legal and social ramifications of peer-to-peer music sharing. https://www.theguardian.com/music/2013/feb/24/napster-music-free-file-sharing

A History of Napster (Lifewire) A timeline of key moments in Napster's development, shutdown, and legacy. https://www.lifewire.com/history-of-napster-2438592

The Revolution and Evolution of Napster A Medium essay examining the broader cultural context around Napster's rise and fall. https://medium.com/@zacharywestonintech/title-the-revolution-and-evolution-of-napster-a-pioneering-tale-in-music-sharing-47ce904489f0

Unraveling the Napster Revolution A Yellowbrick

blog post that frames Napster as a pivotal turning point in the digital music era. https://www.yellowbrick.co/blog/music/unraveling-the-napster-revolution-a-history-of-digital-music-sharing

Key People

Mark Cuban – Wikipedia Biographical information on Mark Cuban, including his business ventures and media influence. https://en.wikipedia.org/wiki/Mark_Cuban

How Mark Cuban Protected His $1 Billion Deal A Yahoo Finance piece on Cuban's financial strategy and protective clauses during the Broadcast.com acquisition. https://finance.yahoo.com/news/heres-mark-cuban-protected-1-183018530.html

Todd Wagner – Wikipedia Background on Todd Wagner's partnership with Cuban and his work at Broadcast.com. https://en.wikipedia.org/wiki/Todd_Wagner

Sean Parker – Wikipedia Biography of Napster co-founder Sean Parker, including his involvement in early tech startups. https://en.wikipedia.org/wiki/Sean_Parker

Internet Radio and Streaming Media

Internet radio – Wikipedia General overview of the development and proliferation of internet-based radio services. https://en.wikipedia.org/wiki/Internet_radio

Streaming media – Wikipedia Definition and historical progression of streaming media as a distribution method. https://en.wikipedia.org/wiki/Streaming_media

Legal and Cultural Impact

Metallica v. Napster, Inc. – Wikipedia Detailed summary of the landmark lawsuit that helped define digital copyright law. https://en.wikipedia.org/wiki/Metallica_v._Napster,_Inc.

Online piracy – Wikipedia Definition, legal context, and ongoing debates around unauthorized distribution of digital content. https://en.wikipedia.org/wiki/Online_piracy

Chapter 6: References

Regulation and Policy in Streaming

Regulating Netflix: Cross-national momentum

This article discusses the global regulatory challenges faced by streaming platforms like Netflix, particularly in relation to local content requirements and cross-national policy frameworks.
https://policyreview.info/articles/news/regulating-netflix-cross-national-momentum/1706

Policy issues on the streaming video horizon

Explores emerging policy challenges for streaming platforms, including taxation, content regulation, and shifts in power between traditional broadcasters and digital services. https://digitalcontentnext.org/blog/2022/10/27/policy-issues-on-the-streaming-video-horizon/

Web3 and Blockchain in Content Creation

How Web3 is redefining content creation

Explains how blockchain and decentralized platforms are changing content creation and monetization models for independent creators and publishers. https://tde.fi/founder-resource/blogs/blockchain-networks/how-web3-is-redefining-content-creation/

The Decline of Cable News and Rise of Streaming

Cable news fact sheet

A Pew Research report showing declining viewership across major cable news networks as more audiences migrate to streaming. https://www.pewresearch.org/journalism/fact-sheet/cable-news/

CNN faces unprecedented crisis after historic ratings collapse

Covers CNN's ratings decline and its strategic shift toward digital-first programming and platform investment. https://www.econotimes.com/CNN-Faces-Unprecedented-Crisis-After-Historic-Ratings-Collapse-Only-61000-Viewers-in-Key-Demo-1694127

CNN invests in digital transformation

A Financial Times article highlighting CNN's major investments in streaming and digital transformation to counter viewer erosion. https://www.ft.com/content/e1368295-b24e-4295-8ff8-7015714286fe

Streaming Statistics and Trends

Traditional TV and streaming viewers in the U.S.
Statista's report showing comparative audience figures for streaming platforms versus traditional broadcast television in the U.S. https://www.statista.com/statistics/1385919/traditional-tv-streaming-viewers-us/

Streaming stats and trends
Forbes outlines current trends in streaming consumption, including growth rates, viewer behavior, and monetization strategies. https://www.forbes.com/home-improvement/internet/streaming-stats/

Streaming trends and stats (GWI)
A GWI blog post examining streaming trends such as increased adoption of ad-supported models and shifting viewer habits. https://blog.gwi.com/marketing/streaming-trends-and-stats/

YouTube numbers: Uncovering the 'ghost town' effect
A TechSpot analysis revealing that most YouTube content receives very little engagement, highlighting the challenges of content saturation and algorithm-driven visibility. https://www.techspot.com/news/106791-youtube-numbers-uncovering-youtube-ghost-town-billions-unwatched.html

Chapter 7: References

RealAudio Introduction RealAudio, introduced by RealNetworks in April 1995, was a groundbreaking technology enabling audio streaming over the internet. https://en.wikipedia.org/wiki/RealAudio

Rob Glaser and RealNetworks Rob Glaser founded RealNetworks after leaving Microsoft, driving innovation in streaming media with products like RealAudio. https://en.wikipedia.org/wiki/Rob_Glaser

Bulletin Board Systems (BBS) Bulletin board systems were early text-based platforms for communication, later succeeded by community-driven streaming like Broadcast.com. https://en.wikipedia.org/wiki/Bulletin_board_system

Broadcast.com Overview Broadcast.com pioneered internet streaming, enabling live sports, corporate events, and more to reach online audiences. https://en.wikipedia.org/wiki/Broadcast.com

Fraunhofer IIS and MP3 Development The Fraunhofer Institute developed the MP3 format, revolutionizing digital audio compression. https://en.wikipedia.org/wiki/Fraunhofer_Society

Yahoo Acquisition of Broadcast.com Yahoo's $5.7 billion acquisition of Broadcast.com was one of the largest deals of the dot-com boom. https://en.wikipedia.org/wiki/Yahoo!

Why Yahoo Failed Yahoo's lack of strategic focus and ineffective integration of acquisitions led to its decline. https://www.feedough.com/why-yahoo-failed, https://tribune.com.pk/story/1153035/yahoos-demise-internet-giants-failure-story-missed-opportunities, https://em360tech.com/tech-article/why-did-yahoo-fail

Windows Media Player and WMA Codec Microsoft introduced Windows Media Player to compete with RealAudio, offering improved stability and performance. https://en.wikipedia.org/wiki/Windows_Media_Player

Codec Evolution From RealAudio to Fraunhofer MP3, codecs were foundational to the early streaming revolution. https://en.wikipedia.org/wiki/Audio_codec

The Concept of Pre-Roll Ads Pre-roll ads, introduced in early streaming to address deep-linking challenges, are now a cornerstone of monetized streaming. https://www.advertisingweek360.com/the-evolution-of-pre-roll-ads/

Internet Bandwidth in the 1990s The high cost and technical limitations of bandwidth were major hurdles for early streaming platforms like Broadcast.com. https://en.wikipedia.org/wiki/Internet_bandwidth

Microsoft and RealNetworks Rivalry RealNetworks accused Microsoft of anti-competitive behavior, highlighting tensions in the early streaming market. https://www.cnet.com/tech/realnetworks-files-antitrust-suit-against-microsoft/

Chapter 8: References

Founding of YouTube YouTube was founded in 2005 by Steve Chen, Chad Hurley, and Jawed Karim, revolutionizing content creation and consumption. https://en.wikipedia.org/wiki/YouTube

First YouTube video: Me at the Zoo The first-ever video uploaded to YouTube was "Me at the Zoo," featuring co-founder Jawed Karim discussing elephants. https://www.youtube.com/watch?v=jNQXAC9IVRw

Google acquires YouTube Google acquired YouTube in November 2006 for $1.65 billion in stock. https://www.nytimes.com/2006/10/09/business/09cnd-deal.html

H.264 video compression technology H.264, also known as Advanced Video Coding (AVC), became the dominant codec for streaming due to its efficiency and quality. https://en.wikipedia.org/wiki/H.264/MPEG-4_AVC

YouTube Partner Program Launched in 2007, the YouTube Partner Program allowed creators to monetize their videos through ad revenue. https://support.google.com/youtube/answer/72902?hl=en

Twitch: Evolution of live streaming Twitch, initially a gaming-focused offshoot of Justin.tv, became a major live-streaming platform with over 140 million monthly active users by 2020. https://en.wikipedia.org/wiki/Twitch_(service)

TikTok's rise to prominence TikTok, launched in 2016, emphasized short-form, mobile-first videos and gained immense popularity globally. https://en.wikipedia.org/wiki/TikTok

Netflix's discontinuation of DVD rentals In 2023, Netflix officially ended its DVD rental service after 25 years. https://www.foxbusiness.com/lifestyle/netflix-officially-terminates-dvd-rental-service-final-mailings-end-of-an-era

The "Adpocalypse" of 2017 Forbes article covering YouTube's monetization policy changes and the resulting impact on creators. https://www.forbes.com/sites/erikkain/2017/09/18/adpocalypse-2017-heres-what-you-need-to-know-about-youtubes-demonetization-troubles/

Broadcast.com and its legacy Broadcast.com paved the way for live-streaming media, later acquired by Yahoo! in 1999 for $5.7 billion. https://en.wikipedia.org/wiki/Broadcast.com

Cultural impact of PewDiePie and other YouTube creators PewDiePie became a global sensation, epitomizing YouTube's potential for individual creators to achieve massive reach. https://en.wikipedia.org/wiki/PewDiePie

Total YouTube video count and upload rates YouTube hosts over 5 billion videos, with more than 360 hours of content uploaded every minute. https://seo.ai/blog/how-many-videos-are-on-youtube

YouTube's global user base in 2025 YouTube is projected to have 2.85 billion users worldwide by 2025. https://www.globalmediainsight.com/blog/youtube-users-statistics/

YouTube's TV dominance YouTube accounts for over 1 billion hours of content viewed daily on TV screens alone. https://www.theverge.com/news/609684/youtube-bigger-tvs-phones-streaming

Chapter 9: References

AV1 codec overview AV1 is a royalty-free video coding format designed for video transmissions over the Internet. https://en.wikipedia.org/wiki/AV1

TCP vs. UDP for video streaming GeeksforGeeks article comparing the benefits and trade-offs of TCP and UDP protocols in the context of video streaming. https://www.geeksforgeeks.org/tcp-vs-udp-for-video-streaming/

Real-Time Transport Protocol (RTP) RTP provides end-to-end delivery services for data with real-time characteristics, such as interactive audio and video. https://www.rfc-editor.org/rfc/rfc3550.html

HLS protocol by Apple HTTP Live Streaming (HLS) delivers media content over HTTP and is widely adopted for adaptive streaming across Apple and third-party devices. https://developer.apple.com/streaming/

Content delivery networks and scalability CDNs like Akamai reduce latency and improve user experiences by caching content closer to users. https://www.akamai.com/

Akamai: The evolution of CDN and cloud services A video overview explaining how Akamai helped shape the development of scalable internet content delivery. https://www.youtube.com/watch?v=9efdk4uzhUg

Netflix's Open Connect CDN Netflix developed Open Connect to optimize video delivery, providing a custom-built CDN tailored for its global user base. https://openconnect.netflix.com/

Cloudflare CDN overview Cloudflare provides a global CDN to improve website performance and secure data transmission across distributed networks. https://www.cloudflare.com/cdn/

How Twitch uses Amazon CloudFront Amazon CloudFront powers Twitch's global live-streaming infrastructure with low-latency, high-speed data delivery. https://aws.amazon.com/cloudfront/

The role of edge computing in streaming Edge computing enhances CDN performance by processing data closer to users, thereby reducing latency and buffering. https://www.ibm.com/cloud/what-is-edge-computing

Dynamic Adaptive Streaming over HTTP (DASH) DASH enables adaptive streaming by dynamically adjusting video quality in response to real-time network conditions. https://www.iso.org/standard/65274.html

The history of Broadcast.com Broadcast.com laid the foundation for live streaming and CDNs, overcoming early internet bandwidth limitations. https://en.wikipedia.org/wiki/Broadcast.com

Microsoft Azure Content Delivery Network Azure CDN integrates global scaling with dynamic content acceleration, designed for streaming media applications. https://azure.microsoft.com/en-us/products/cdn/

Chapter 10: References

Akamai CDN Overview Details how CDNs like Akamai reduce latency and improve streaming scalability. https://www.akamai.com/

Amazon Kindle 1984 Deletion Reports on Amazon's 2009 remote deletion of Kindle copies of *1984*, sparking ownership debates. https://www.nytimes.com/2009/07/18/technology/companies/18amazon.html

Amazon Prime Video Ownership Lawsuit Discusses Amazon's 2020 court argument that customers don't own purchased digital content. https://news.bloomberglaw.com/litigation/amazon-cant-dislodge-purchased-digital-content-deception-suit

Amazon Removing Kindle USB Transfer Covers Amazon's removal of Kindle download and USB transfer options in 2024. https://www.theverge.com/2024/10/31/24189787/amazon-kindle-download-transfer-via-usb-removal

Apple App Store Ecosystem Growth Apple's report on the App Store's role in digital media distribution. https://www.apple.com/newsroom/pdfs/the-continued-growth-and-resilience-of-apples-app-store-ecosystem.pdf

Apple Streaming with HLS Explains Apple's HTTP Live Streaming (HLS) protocol for adaptive streaming. https://developer.apple.com/streaming/

Apple TV+ Android Expansion Covers Apple TV+ becoming available on Android devices as of 2025. https://www.theaustralian.com.au/business/technology/apple-tv-finally-comes-to-android-phones-as-the-tech-titan-woos-new-customers/news-story/cf043a11bc9d5328560b183c5e9ff45a

Apple TV+ Free Weekend Discusses Apple TV+'s market share and promotional strategy. https://www.sfgate.com/streaming/article/apple-tv-plus-free-weekend-20013725.php

AT&T DSL and Fiber History A historical overview of AT&T's broadband expansion and policy context. https://progressivepolicy.org/

wp-content/uploads/2014/03/2014.03-Ehrlich_A-Brief-History-of-Internet-Regulation.pdf

AT&T Fiber Expansion Outlines AT&T's recent fiber initiatives and network strategy.
https://about.att.com/story/2024/att-fiber-expansion.html

AT&T Integrated Gateway Launch Details the launch of AT&T's 2024 fiber and 5G integrated gateway for business use. https://about.att.com/story/2024/integrated-gateway.html

AT&T Network Upgrades Summarizes AT&T's 2023 plans for upgrading its fiber and 5G infrastructure. https://about.att.com/blogs/2023/upgrading-the-network.html

AV1 Codec Overview Wikipedia entry on AV1, a royalty-free video codec for efficient internet streaming. https://en.wikipedia.org/wiki/AV1

Calix SmartTown Wi-Fi Describes city-scale Wi-Fi networks to enhance digital connectivity.
https://www.calix.com/products/smartlife/smarttown.html

Comcast and Time Warner Cable History Chronicles the competitive history of Comcast and Time Warner Cable. https://www.cnbc.com/2014/02/13/history-of-comcast-and-time-warner-cable.html

Deloitte Digital Media Trends Explores how social media is shaping entertainment consumption habits. https://www2.deloitte.com/us/en/insights/industry/technology/digital-media-trends-consumption-habits-survey/2024/online-creators-and-the-impact-of-social-media-on-entertainment.html

Digital Media Players on eBay Wikipedia page covering digital media players and platforms like eBay as distribution tools. https://en.wikipedia.org/wiki/Digital_media_player

DOCSIS 4.0 Technology Explains DOCSIS 4.0 as a standard for competitive high-speed cable broadband. https://www.cablelabs.com/technologies/docsis-4-0-technology

eBay Enterprise Streaming Covers eBay's internal use of an enterprise-grade, YouTube-style platform. https://www.streamingmedia.com/Articles/Editorial/Featured-Articles/eBay-Embraces-Enterprise-YouTube-97146.aspx

eBay's Internet Influence Explores eBay's early influence on digital commerce and engagement. https://www.axios.com/2018/06/24/ebay-shaped-modern-internet-google-facebook

eBay Love Island Campaign Details eBay's branded streaming campaign integrated with *Love Island*. https://www.thecurrent.com/tv-ctv-ebay-streaming-soulmate-love-island

Edge Computing in Streaming IBM explains edge computing's role in improving performance and reducing latency in streaming. https://www.ibm.com/cloud/what-is-edge-computing

Ericsson 5G Impact Outlines the key contributions of 5G to streaming media and mobile infrastructure. https://www.ericsson.com/en/5g

FCC Telecommunications Act of 1996 Official overview of the landmark U.S. policy affecting broadband competition. https://www.fcc.gov/general/telecommunications-act-1996

Fiber Optic Communication Evolution Wikipedia entry describing the evolution of fiber optics in broadband. https://en.wikipedia.org/wiki/Fiber-optic_communication

Fiber Optic History Timeline Covers the commercial timeline of fiber optic deployment. https://www.ecmag.com/magazine/articles/article-detail/fiber-optic-history-timeline

Forbes Mobile Connectivity Evolution Analyzes the transition from 3G to 5G and its impact on streaming. https://www.forbes.com/councils/forbestechcouncil/2020/12/18/3g-4g-5g-and-beyond-the-quest-for-mobile-connectivity-and-speed/

Harmonic Video Trends A review of sustainability and bandwidth demands in the video space. https://www.harmonicinc.com/insights/blog/video-trends/

History of Fiber Internet Race Communications blog on the origin and expansion of fiber broadband. https://race.com/blog/history-of-fiber-internet/

Impact of Live Streaming on Marketing Examines how real-time video transformed social media strategies. https://www.nobledesktop.com/learn/social-media-marketing/the-impact-of-live-streaming-on-social-media-marketing

Inoxoft Streaming Features Lists essential modern features in successful streaming platforms. https://inoxoft.com/blog/essential-features-for-streaming-platforms/ **Kadence Live Streaming Impact** Details how live streaming is transforming media consumption habits. https://kadence.com/en-us/how-live-streaming-is-reshaping-media-consumption

Matter Smart Home Standard Explains Matter, the unified smart home protocol supported by major tech companies. https://www.wired.com/story/what-is-matter/

Music Streaming Economic Impact A DIMA study showing the job creation and revenue impact of music streaming. https://dima.org/news-and-resources/new-study-highlights-music-streaming-outsized-impact-us-economy/

Muvi OTT Advancements Covers innovations in over-the-top (OTT) streaming technology. https://www.muvi.com/blogs/the-technological-advancements-in-ott-platforms/

Netflix Economic Impact in South Korea Highlights Netflix's influence on South Korean media and content exports. https://www.fdiintelligence.com/content/locations/global/fresh-content-how-streaming-platforms-are-changing-the-geography-of-media-investment-80490

Netflix Open Connect CDN Overview of Netflix's proprietary CDN used for efficient video delivery. https://openconnect.netflix.com/

Network Convergence Wikipedia article explaining how telecom and media networks have merged. https://en.wikipedia.org/wiki/Network_convergence

Online Video Platform Statistics Market data on online video usage and streaming by device. https://scoop.market.us/online-video-platform-statistics/

OTT Messaging Services Explores messaging services that operate independently of traditional carriers. https://en.wikipedia.org/wiki/Over-the-top_media_service

Qualcomm 4G LTE Role Highlights Qualcomm's foundational role in LTE adoption. https://www.qualcomm.com/research/5g/4g

Qualcomm 5G Overview Outlines 5G's core technologies and streaming implications. https://www.qualcomm.com/5g/what-is-5g

Qualcomm 5G Technology Focuses on Qualcomm's hardware for next-generation networks. https://www.qualcomm.com/products/technology/5g

Real-Time Transport Protocol (RTP) Official RFC describing RTP's role in real-time audio and video transport. https://www.rfc-editor.org/rfc/rfc3550.html

Satellite Internet and Digital Divide BBC article on satellite technology's role in closing connectivity gaps. https://www.bbc.com/news/technology-56740488

Starlink Overview Wikipedia page detailing Starlink's mission, rollout, and global coverage. https://en.wikipedia.org/wiki/Starlink

Streaming Changed the Music Industry Examines how streaming services disrupted traditional music business models. https://aimm.edu/blog/how-streaming-changed-the-music-industry

T-Mobile and SpaceX Partnership Details the collaboration between T-Mobile and Starlink for direct satellite-to-device service. https://www.t-mobile.com/news/un-carrier/t-mobile-takes-coverage-above-and-beyond-with-spacex

Telecom Act Competition Analysis Brookings Institution review of the effectiveness of the 1996 Act. https://www.brookings.edu/articles/was-the-1996-telecommunications-act-successful-in-promoting-competition/

Telecom Consolidation Concerns Law journal article analyzing industry concentration trends. https://www.repository.law.indiana.edu/cgi/viewcontent.cgi?article=1445&context=fclj

Text Messaging Advantages Highlights benefits of OTT messaging platforms for customer engagement. https://www.textmagic.com/blog/10-online-text-messaging-advantages-you-havent-considered

TikTok's Music Industry Impact Explores how TikTok drives music discovery and viral trends. https://www.businessinsider.com/how-tiktok-is-changing-music-industry

Time Warner Cable History Wikipedia entry outlining TWC's corporate evolution and broadband role. https://en.wikipedia.org/wiki/Time_Warner_Cable

UDP in Streaming Cloudflare's overview of how UDP enables low-latency streaming. https://www.cloudflare.com/learning/network-layer/what-is-udp/

Verizon FiOS Launch Press release detailing the 2005 launch and impact of Verizon FiOS. https://www.verizon.com/about/news/press-releases/two-years-after-launch-verizon-fios-tv-continues-break-new-ground-subscription-tv-market

Verizon High-Speed Broadband Describes Verizon's 5G and fiber offerings for broadband customers. https://www.verizon.com/about/our-company/high-speed-broadband

Wi-Fi 6 and 6E Overview Wi-Fi Alliance's overview of the latest wireless standards. https://www.wi-fi.org/discover-wi-fi/wi-fi-certified-6

Wi-Fi Standards Evolution Explains differences among major Wi-Fi protocols like 802.11n and 802.11ac. https://www.cbtnuggets.com/blog/technology/networking/when-to-use-802-11-a-b-g-b-nc-wifi-standards

Yellowbrick Streaming Trends Forecasts key technology shifts in the streaming media industry. https://www.yellowbrick.co/blog/entertainment/the-future-of-streaming-technology-key-trends-and-innovations

Starlink and T-Mobile FCC Approval Reports on the FCC's 2024 approval of Starlink's direct-to-device service in partnership with T-Mobile. https://www.theverge.com/2024/11/27/24307394/starlink-spacex-tmobile-direct-to-cell-satellite-fcc-approval

Chapter 11: References

Manage Live Stream Latency – YouTube Official YouTube Help page detailing how YouTube Live manages latency using Normal,

Low, and Ultra-low latency modes for HLS and DASH streaming….
https://support.google.com/youtube/answer/7444635

Twitch: Low Latency, High Reach Twitch blog post outlining their low-latency streaming network, emphasizing Low-Latency HLS and real-time engagement benefits. https://blog.twitch.tv/en/2021/10/25/low-latency-high-reach-creating-an-unparalleled-live-video-streaming-network-at-twitch/

Twitch Low-Latency Video Help Guide Twitch Help article explaining how to enable low-latency mode, highlighting benefits and trade-offs such as buffering on slower connections. https://help.twitch.tv/s/article/low-latency-video

What Is WebRTC – Wowza Wowza blog post explaining WebRTC's use for real-time communications such as Zoom and live events, with emphasis on latency. https://www.wowza.com/blog/what-is-webrtc

Scaling WebRTC Applications – LiveSwitch LiveSwitch blog post discussing scaling WebRTC via SFUs, relevant to Zoom and real-time applications requiring low latency. https://www.liveswitch.io/blog/how-to-successfully-scale-your-webrtc-application-in-2021

MPEG-DASH Overview Wikipedia entry on Dynamic Adaptive Streaming over HTTP (DASH), detailing its use in services like Netflix and its focus on stream stability over latency. https://en.wikipedia.org/wiki/Dynamic_Adaptive_Streaming_over_HTTP

DAZN & LTN Partnership for Low-Latency Streams Sports Video Group article on DAZN and LTN launching 11 low-latency streaming channels (sub-300ms) to betting venues. https://www.sportsvideo.org/2024/12/10/dazn-ltn-expand-partnership-to-deliver-11-low-latency-live-streaming-channels-to-betting-locations/

Gcore CDN Technology for Ultra-Low Latency SportsPro article examining how Gcore enables ultra-low-latency streaming and scalability challenges for platforms like ESPN+ and DAZN. https://www.sportspro.com/insights/gcore-cdn-tech-streaming-ultra-low-latency/

A Better Way to Scale WebRTC – Phenix Phenix blog post discussing their approach to scaling WebRTC for large audiences in sports betting and interactive use cases. https://blog.phenixrts.com/a-better-way-to-scale-webrtc

Streaming Protocols and Latency Comparison – Wowza Wowza article comparing HLS, DASH, RTP, and WebRTC protocols and their trade-offs in latency, scale, and application use. https://www.wowza.com/blog/streaming-protocols-latency

WebRTC vs. HLS – Gumlet Gumlet article comparing WebRTC and HLS for live streaming, focusing on the latency vs. compatibility decision https://www.gumlet.com/learn/webrtc-vs-hls/

LL-HLS, CMAF, and WebRTC – Cloudinary Cloudinary guide comparing streaming options, evaluating latency and infrastructure costs of low-latency approaches. https://cloudinary.com/guides/live-streaming-video/low-latency-hls-ll-hls-cmaf-and-webrtc-which-is-best

Edge Computing and Streaming – Muvi Muvi blog post describing how edge computing lowers latency for services like Twitch and ESPN+ by localizing data processing. https://www.muvi.com/blogs/role-of-edge-computing-in-video-streaming/

AI Reshaping Entertainment – Forbes Forbes article by Neil Sahota on how AI is improving real-time video quality, adjusting compression, and forecasting congestion. https://www.forbes.com/sites/neilsahota/2024/03/18/streaming-into-the-future-how-ai-is-reshaping-entertainment/

Edge Computing in Streaming – Verizon Business Verizon article exploring how 5G and edge computing enhance audio and video performance for real-time streaming platforms. https://www.verizon.com/business/resources/articles/s/how-edge-computing-can-help-improve-audio-and-video-streaming-technology/

Chapter 12: References

ESPN Flagship Direct-to-Consumer Service Provides information on ESPN's upcoming streaming service, expected to launch in late summer 2025, offering content from both ESPN's linear channels and ESPN+. https://en.wikipedia.org/wiki/ESPN_Flagship

Max (Formerly HBO Max) Sports Streaming Tier Wikipedia entry outlining Max's expansion into sports with a dedicated streaming tier. https://en.wikipedia.org/wiki/Max_(streaming_service)

Max Streaming Service's Bleacher Report Sports Add-On Covers Max's launch of a sports-focused package featuring live events and exclusive content. https://en.wikipedia.org/wiki/Max_%28streaming_service%29

Netflix Considering Bid for Formula 1 TV Rights Explores Netflix's interest in acquiring U.S. Formula 1 broadcasting rights, marking its move into live sports. https://www.thetimes.co.uk/article/netflix-considering-bid-for-formula-1-tv-rights-9xd50qlm6

DAZN's Challenges in Profitability Due to High Rights Costs Financial Times article examining how rights costs affect DAZN's business model. https://www.ft.com/content/4aff9714-f70f-4ce1-ac8d-f2fcf682a434

NFL Media Rights Historical Overview Wikipedia entry

providing a historical look at the NFL's U.S. broadcast arrangements.
https://en.wikipedia.org/wiki/NFL_on_American_television

History of NFL Broadcasting Details the evolution of the
NFL's presence on American television. https://en.wikipedia.org/wiki/
History_of_the_NFL_on_television

**Fox Sports' Strategic Use of Tom Brady in NFL
Broadcasts** Article detailing how Fox plans to leverage
Tom Brady's brand to strengthen NFL viewership amid
streaming competition. https://talksport.com/nfl/2984347/
fox-sports-streaming-service-tom-brady-2025-football-season/

NBA's 2025–26 Media Rights Deal Details Breaks down
the NBA's new streaming and broadcast strategy entering the
2025–26 season. https://www.sportsbusinessjournal.com/Articles/2024/07/10/
nba-media-rights-streaming-how-to-watch

NASCAR's New Broadcasting Agreements Wikipedia entry
covering NASCAR's 2025–2031 media deals with Fox, NBC,
Amazon Prime Video, and TNT Sports. https://en.wikipedia.org/wiki/
Sports_broadcasting_contracts_in_the_United_States

MLB and ESPN Agree to End Long-Term TV Deal Wall
Street Journal article highlighting the end of MLB's long-term
broadcasting relationship with ESPN. https://www.wsj.com/business/media/
espn-mlb-agree-to-end-long-term-tv-deal-1e0dde11

**TKO Group 'Furious' with ESPN After UFC 313 Streaming
Issues** Coverage of streaming problems during UFC 313 on ESPN+
and their impact on future licensing negotiations. https://nypost.
com/2025/03/09/sports/tko-brass-furious-with-espn-after-ufc-313-streaming-issues/

**Strategic Importance of Pay-Per-View Models in Live
Sports** Outlines how pay-per-view remains a critical monetization
strategy for premium sports events. https://blog.onstreammedia.com/
ppv-live-streaming-services

Pay-Per-View Piracy Crisis and Consumer Pushback
BoxingScene article examining the growing crisis of PPV
piracy and consumer discontent. https://www.boxingscene.com/
pay-per-view-piracy-reaching-crisis-point--184223

**Digital Piracy's Impact on Live Sports Streaming
Revenue** IPWatchdog analysis of how piracy threatens the
economics of live sports streaming. https://ipwatchdog.com/2025/01/23/
ip-octagon-saving-live-sports-digital-piracy/id=185316/

Impact of Cord-Cutting on Regional Sports Networks
Wikipedia article discussing the decline of regional sports networks
and the shift to streaming. https://en.wikipedia.org/wiki/Regional_sports_network

Scripps Sports' Expansion into Sports Broadcasting Wikipedia entry covering Scripps' aggressive move into local and national sports broadcasting. https://en.wikipedia.org/wiki/Scripps_Sports

Chapter 13: References

Disney's Snow White (2024) Controversies

The Sun article on backlash surrounding the portrayal of dwarfs in Disney's remake. https://www.thesun.co.uk/tv/33844272/disney-woke-snow-white-dwarfs-feud/

New York Post coverage of Rachel Zegler's tribute to the original Snow White actress, which drew public criticism. https://nypost.com/2025/03/18/entertainment/rachel-zegler-sparks-backlash-with-snow-white-tribute-to-original-actress/

The Sun article warning about "woke" messaging in Disney's Snow White remake. https://www.thesun.co.uk/tv/33917367/disney-snow-white-woke-warnings-trees/

New York Post report on Rachel Zegler's appearance in Spain while co-star Gal Gadot was absent. https://nypost.com/2025/03/13/entertainment/rachel-zegler-attends-snow-white-spain-event-gal-gadot-skips/

Vanity Fair's comprehensive review of the controversies surrounding Disney's Snow White production. https://www.vanityfair.com/hollywood/story/the-curse-of-snow-white-all-the-drama-around-disneys-remake-explained

New York Post article on criticism from "Little Women: LA" cast over Disney's use of CGI dwarfs. https://nypost.com/2025/03/13/entertainment/snow-white-slammed-by-little-women-la-star-over-cgi-dwarves-we-were-shut-out/

New York Post on Rachel Zegler defending her vocal performance in response to fan criticism. https://nypost.com/2025/03/14/entertainment/rachel-zegler-bites-back-after-fan-claims-shes-not-a-soprano-in-disneys-snow-white/

Marvel's Secret Invasion Opening Credits (2023) AI Controversy

Entertainment Weekly article on fan and industry reaction to

Marvel's AI-generated title sequence.
https://ew.com/tv/secret-invasion-marvel-ai-generated-intro-controversy/

Polygon article on the controversy surrounding the use of AI in Secret Invasion's credits. https://www.polygon.com/23767640/ai-mcu-secret-invasion-opening-credits

Artnet News report on criticism of Marvel's use of generative AI. https://news.artnet.com/art-world/marvel-secret-invasion-ai-opening-credits-2325932

Times of India coverage of Marvel's defense of AI usage in Secret Invasion's visuals. https://timesofindia.indiatimes.com/gadgets-news/marvel-studios-secret-invasion-faces-backlash-over-ai-generated-scenes-heres-what-the-creators-has-to-say/articleshow/101222684.cms

Cartoon Brew article on the creative industry's response to Marvel's AI credit sequence. https://www.cartoonbrew.com/series/marvel-secret-invasion-ai-opening-credits-229988.html

IGN report on the studio's defense that no artists lost jobs due to the AI-created intro. https://www.ign.com/articles/studio-behind-secret-invasions-ai-intro-addresses-backlash-no-artists-jobs-were-replaced

Indiana Jones and the Dial of Destiny (2023) Financial Performance

Box Office Mojo data on the franchise's earnings. https://www.boxofficemojo.com/franchise/fr3209137925/

Collider interview with director James Mangold about audience reception. https://collider.com/indiana-jones-and-the-dial-of-destiny-james-mangold-reacts-audience-reception/

Screen Rant article analyzing the film's box office profitability. https://screenrant.com/indiana-jones-5-movie-box-office-profit-loss/

AI Integration in Sports Broadcasting

AWS blog on the Bundesliga's use of generative AI for live sports commentary. https://aws.amazon.com/blogs/media/revolutionizing-fan-engagementcer-bundesliga-generative-ai-powered-live-commentary/

Rocking Robots article on AI-powered live sports narration. https://www.rockingrobots.com/ai-provides-live-sports-commentary/

ArXiv research paper on AI commentary for sports broadcasts. https://arxiv.org/abs/2402.15514

Reddit post showcasing a working prototype of an AI

live sports commentator. https://www.reddit.com/r/MachineLearning/comments/1fnry1x/p_i_built_a_live_ai_sports_commentator_that_can/

YouTube demo of AI-powered sports commentary in real time. https://www.youtube.com/watch?v=u56K4dL20gA

NBA AI-Driven Personalized Highlights

YouTube clip demonstrating real-time highlight generation by AI. https://www.youtube.com/watch?v=3NKjoKvFwAQ

Sports Video Group article on NBA Digital's automated video clipping system for real-time content. https://www.sportsvideo.org/2015/12/07/nba-digitals-new-automated-video-clipping-solution-generates-real-time-customizable-content/

Chapter 14: References

Sports Fans Using Phones at Events 82% of fans use mobile devices at live sporting events, but connectivity remains a significant challenge. https://www.mediapost.com/publications/article/373404/82-of-fans-use-mobile-devices-at-live-sporting-ev.html

Distributed Antenna Systems (DAS) Overview DAS systems improve cellular coverage in large venues like stadiums by distributing antenna signals across the space. https://www.rcrwireless.com/2022/05/10/das-systems-explained

Private 5G in Stadiums How private 5G networks are revolutionizing connectivity in sports stadiums and enhancing the fan experience. https://www.verizon.com/business/resources/articles/5g-stadiums/

Challenges of Public WiFi in Stadiums The challenges of implementing public WiFi networks in large venues, including congestion and coverage gaps. https://www.networkworld.com/article/3293989/stadium-wifi-networks.html

Edge Computing and Sports Venues Edge computing reduces latency and enhances the in-stadium digital experience by processing data closer to users. https://www.ibm.com/cloud/what-is-edge-computing

5G at SoFi Stadium SoFi Stadium integrates 5G technology to deliver ultra-low latency and high-speed connectivity for fans. https://www.sofistadium.com/5g-network-integration/

In-Stadium Streaming Latency Issues Latency in live streaming continues to be a major issue for fans inside stadiums. https://

www.streamingmedia.com/Articles/Editorial/Featured-Articles/Latency-in-Live-Streaming-147218.aspx

Fan Expectations and Technology Sports fans increasingly expect personalized, connected experiences driven by technology. https://explodingtopics.com/blog/sports-tech-trends

Stadium WiFi and Cellular Upgrades Analyzing the impact of network upgrades on the in-stadium experience for fans. https://www.bleedingcool.com/stadium-wifi-connectivity-upgrades

Private 5G Networks Explained Private 5G offers dedicated high-speed connectivity for enterprises, including sports venues. https://www.qualcomm.com/research/private-5g-networks

Latency Challenges in Live Sports Betting Low latency is critical for live sports betting, but achieving it in crowded stadiums remains a challenge. https://www.sportsbusinessjournal.com/Articles/2023/06/10/live-betting-latency

Public WiFi Security Risks Public WiFi networks in stadiums expose users to significant security risks, including data theft. https://www.csoonline.com/article/public-wifi-risks.html

Hybrid Connectivity Solutions for Venues Combining WiFi, 5G, and DAS for robust stadium connectivity. https://www.arubanetworks.com/solutions/large-venues/

Technology and Stadium Architecture Challenges Designing connected stadiums for better fan experiences and streamlined operations. https://www.networkcomputing.com/network-management/designing-connected-stadiums-for-better-fan-experiences-and-streamlined-operations

Chapter 15: References

82% of Fans Use Phones at Live Sports Events 82% of fans actively use mobile devices during live sports events, highlighting the need for better connectivity and streaming solutions. https://www.mediapost.com/publications/article/373404/82-of-fans-use-mobile-devices-at-live-sporting-ev.html

5G and Stadium Connectivity How 5G networks are transforming stadium connectivity to enhance the fan experience. https://www.verizon.com/business/resources/articles/5g-stadiums/

Edge Computing for Real-Time Streaming Edge computing reduces latency by processing data closer to end users, making real-time streaming viable at scale. https://www.ibm.com/cloud/what-is-edge-computing

Low-Latency Streaming Challenges and Solutions Latency in live streaming remains a significant challenge, but new technologies are emerging to tackle the issue. https://www.streamingmedia.com/Articles/Editorial/Featured-Articles/How-to-Reduce-Latency-in-Live-Streaming-146034.aspx

Geo-Fencing Explained Geo-fencing technology allows services to tailor content and features to users within specific locations. https://www.geofencing.com/how-it-works/

Private 5G Networks in Sports Venues Private 5G networks offer stadiums the ability to deliver reliable, high-speed, low-latency connectivity to fans. https://www.qualcomm.com/research/private-5g-networks

The Role of Real-Time Commentary in Sports Engagement How real-time, multilingual commentary enhances fan engagement and broadens global sports audiences. https://www.sportsbusinessjournal.com/Articles/2023/06/15/multilingual-commentary-in-sports

Latency and Live Sports Challenges The latency debate in live sports: consistency vs. speed. https://dolby.io/blog/the-latency-debate-in-live-sports-consistency-vs-speed/

SoFi Stadium's 5G Integration SoFi Stadium has integrated 5G to enable advanced connectivity features for fans during events. https://www.sofistadium.com/5g-network-integration/

Proprietary Low-Latency Streaming Solutions Exploring proprietary technologies that minimize latency for real-time streaming applications. https://www.streamingmedia.com/Articles/Editorial/Featured-Articles/Low-Latency-Streaming-Solutions-147218.aspx

Monetizing Fan Engagement: Turning Passion into Revenue Explores strategies for leveraging fan engagement to create sustainable revenue models. https://www.fancompass.com/blog/2024/8/21/monetizing-fan-engagement-turning-passion-into-revenue

Maximizing ROI in Fan Engagement Discusses key pillars for driving ROI through fan-focused technology and engagement platforms. https://www.sportstechgroup.org/paid-for-content/maximizing-roi-in-fan-engagement-the-four-pillars-of-effective-realtime-experience-platforms/

Marketing Business Grew Big by Thinking Small Highlights how businesses are tailoring small-scale innovations to make significant impacts in sports. https://www.thetimes.co.uk/article/marketing-business-grew-big-by-thinking-small-in-world-of-sport-enterprise-network-pz7m2bjsb

People Expect an Experience at Live Events Explores the growing importance of experiential elements in live sports events. https://www.thetimes.co.uk/article/whether-in-the-stands-or-hospitality-people-expect-an-experience-jwgpm3532

Washington Commanders: Business is Booming Discusses how branded merchandise and innovative strategies are driving revenue growth for sports teams. https://www.hogshaven.com/2024/11/10/24290942/ daily-slop-10-nov-24-business-is-booming-for-the-washington-commanders

Fan Expectations and Technology Trends in Sports A deep dive into how technology is reshaping fan expectations and engagement strategies. https://explodingtopics.com/blog/sports-tech-trends

Chapter 16: References

Sports Technology Market Size and Growth The global sports technology market is expected to reach $87.5 billion by 2028, driven by AI, AR, and real-time streaming advancements. https://www.grandviewresearch.com/industry-analysis/sports-technology-market

5G Networks and Real-Time Streaming in Sports Private 5G networks are transforming connectivity in sports venues, enabling low-latency streaming and interactive experiences. https://www.qualcomm.com/research/private-5g-networks

AI in Sports Analytics and Streaming Artificial intelligence is reshaping sports through predictive analytics, automated content creation, and personalized experiences. https://www.ibm.com/topics/sports-analytics

Latency in Live Sports Streaming Low-latency streaming is critical for live sports, enabling real-time interaction, betting, and enhanced fan experiences. https://www.streamingmedia.com/Articles/Editorial/Featured-Articles/How-to-Reduce-Latency-in-Live-Streaming-146034.aspx

The Rise of Personalized Sports Content AI-driven personalization is enabling fans to customize their sports streaming experiences like never before. https://www.techradar.com/news/personalized-sports-content-streaming

AI-Powered Predictive Analytics in Sports Predictive analytics powered by AI helps teams, broadcasters, and fans anticipate outcomes and analyze performances in real time. https://www.nytimes.com/2023/12/15/sports/ai-predictive-analytics.html

The Economics of Sports Streaming Analyzing the financial opportunities and challenges in the rapidly evolving sports streaming market. https://www.mckinsey.com/industries/media-and-entertainment/our-insights/economics-of-sports-streaming

5G Revolutionizing Mobile Streaming for Fans 5G technology is setting the stage for ultra-low-latency, high-definition mobile streaming, revolutionizing fan engagement. https://www.ericsson.com/en/5g/sports-streaming